S0-CMZ-113

Applied Mathematical Sciences | Volume 3

J. Hale

Functional Differential Equations

With 15 Illustrations

Springer-Verlag New York · Heidelberg · Berlin
1971

Jack K. Hale
Division of Applied Mathematics
Brown University
Providence, Rhode Island

Library of Congress Catalog Card Number 73-149142
Printed in the United States of America

ISBN 0-387-90023-3 Springer-Verlag New York • Heidelberg • Berlin
ISBN 3-540-90023-3 Springer-Verlag Berlin • Heidelberg • New York

PREFACE

It is hoped that these notes will serve as an introduction to the subject of functional differential equations. The topics are very selective and represent only one particular viewpoint. Complementary material dealing with extensions of closely related topics are given in the notes at the end. A short bibliography is appended as source material for further study.

The author is very grateful to the Mathematics Department at UCLA for having extended the invitation to give a series of lectures on functional differential equations during the Applied Mathematics Year, 1968-1969. The extreme interest and sincere criticism of the members of the audience were a constant source of inspiration in the preparation of the lectures as well as the notes. Except for Sections 6, 32, 33, 34 and some other minor modifications, the notes represent the material covered in two quarters at UCLA. The author wishes to thank Katherine McDougall and Sandra Spinacci for their excellent preparation of the text. The author is also indebted to Eleanor Addison for her work on the drawings and to Dr. H. T. Banks for his careful proofreading of this material.

Jack K. Hale
Providence
March 4, 1971

TABLE OF CONTENTS

Applied Mathematical Sciences | Volume 3

1. INTRODUCTION

In the applications, the future behavior of many phenomena are assumed to be described by the solutions of an ordinary differential equation. Implicit in this assumption is that the future behavior is uniquely determined by the present and independent of the past. In differential difference equations (DDE), or more generally functional differential equations (FDE), the past exerts its influence in a significant manner upon the future. Many models under scruitiny are better represented by FDE, than by ordinary differential equations.

DDE and FDE were first encountered in the late eighteenth century by the Bernoulli's, Laplace and Condorcet. However, very little was accomplished during the nineteenth century and the early part of the twentieth century. During the last forty years and especially the last twenty, the subject has been and is continuing to be investigated at a very rapid pace. The impetus has mainly been due to developments in the theory of control, mathematical biology, mathematical economics and the theory of systems which communicate through lossless channels.

In this introductory section, we indicate by means of examples (with references to their origin) the diversity of FDE.

Minorsky [1, Ch. 21] was one of the first investigators of modern times to study the differential-difference equation

$$\dot{x}(t) = F(t,x(t),x(t-r))$$

and its effect on simple feedback control systems in which the communication time cannot be neglected.

Lord Cherwell (see Wright [1]) has encountered the differential-difference equation

$$\dot{x}(t) = -\alpha x(t-1)(1+x(t))$$

in his study of the distribution of primes. Variants of this equation have also been used as models in the theory of growth (see Cunningham [1]).

Volterra [1] in his study of predator-prey models studied the integro-differential equations

$$\dot{N}_1(t) = [\varepsilon_1 - \gamma_1 N_2(t) - \int_{-r}^{0} F_1(-\theta)N_2(t+\theta)d\theta]N_1(t)$$

$$\dot{N}_2(t) = [-\varepsilon_2 + \gamma_2 N_1(t) + \int_{-r}^{0} F_2(-\theta)N_1(t+\theta)d\theta]N_2(t),$$

where N_1, N_2 are the number of prey and predators, respectively.

Wangersky and Cunningham [1] have also used the equations

$$\dot{x}(t) = \alpha(x(t))[\frac{m-x(t)}{m}] - bx(t)y(t)$$

$$\dot{y}(t) = -\beta y(t) + cx(t-r)y(t-r)$$

for similar models.

The equation

$$\dot{x}(t) = -\int_{t-r}^{t} a(t-u)g(x(u))du$$

was encountered by Ergen [1] in the theory of a circulating fuel nuclear reactor and has been studied extensively by Levin and Nohel [1]. In this model, x is the neutron density. It is also a good model in one dimensional viscoelasticity in which x is the strain and a is the relaxation function.

In the theory of control, Krasovskii [1] has studied extensively the system

$$\dot{x}(t) = P(t)x(t) + B(t)u(t)$$

$$y(t) = Q(t)x(t)$$

$$\dot{u}(t) = \int_{-r}^{0} [d_\theta \eta(t,\theta)]y(t+\theta) + \int_{-r}^{0} [d_\theta \mu(t,\theta)]u(t+\theta).$$

In theory of lossless transmission lines, Miranker [1] and Brayton [1] have encountered the equation

$$\dot{v}(t) = \alpha\dot{v}(t-r) - \beta v(t) - \alpha\gamma v(t-r) + F(v(t),v(t-r))$$

where α,β,γ are constants.

In his study of vibrating masses attached to an elastic bar, Rubanik [1] considered the equations

$$\ddot{x}(t) + \omega_1^2 x(t) = \varepsilon f_1(x(t),\dot{x}(t),y(t),\dot{y}(t)) + \gamma_1\ddot{y}(t-r)$$
$$\ddot{y}(t) + \omega_2^2 y(t) = \varepsilon f_2(x(t),\dot{x}(t),y(t),\dot{y}(t)) + \gamma_2\ddot{x}(t-r).$$

In studying the collision problem in electrodynamics, Driver [1] encountered systems of the type

$$\dot{x}(t) = f_1(t,x(t),x(g(t))) + f_2(t,x(t),x(g(t)))\dot{x}(g(t)), \quad g(t) \le t.$$

El'sgol'tz [1] and Hughes [1] have considered the following variational problem: minimize

$$V(x) = \int_0^1 F(t,x(t),x(t-r),\dot{x}(t),\dot{x}(t-r))dt$$

over some class of functions x. Generally, the Euler equations are of the form

$$\ddot{x}(t) = f(t,x(t),x(t-r),\dot{x}(t),\dot{x}(t-r),\ddot{x}(t-r)).$$

In the slowing down of neutrons in a nuclear reactor the equation

$$x(t) = \int_t^{t+1} k(s)x(s)ds$$

or

$$\dot{x}(t) = k(t+1)x(t+1) - k(t)x(t)$$

seems to play an important role (see Slater and Wilf [1]).

As we see from the above, there are many different types of equations that occur in the applications - some which depend only upon the past state, some which depend upon the past state as well as the rate of change of the past state and some which depend upon the future. The solutions behave differently for each of these types of equations. To recognize some of the difficulties, let us discuss in an intuitive manner some very simple examples.

Consider first the linear <u>retarded equation</u>

$$\dot{x}(t) = -x(t-r), \quad r > 0. \tag{1.1}$$

What is the minimum amount of data that is necessary for (1.1) to define a function for $t \geq 0$? A moment of reflection indicates that we must specify a function on the interval $[-r,0]$. If φ is a given continuous function defined on $[-r,0]$, then there is only one function $x(t)$ defined on $[-r,\infty)$ which coincides with φ on $[-r,0]$ and satisfies (1.1) for $t > 0$. In fact, if x is such a function, then it must satisfy

$$x(t) = \varphi(0) - \int_0^t x(s-r)ds, \quad t \geq 0 \tag{1.2}$$

and in particular,

$$x(t) = \varphi(0) - \int_0^t \varphi(s-r)ds, \quad 0 \leq t \leq r.$$

This latter equation uniquely defines x on $[0,r]$. Once x is known on $[0,r]$, then (1.2) uniquely defines x on $[r,2r]$, etc.

The following observations about (1.1) are important:

(I) For any continuous function φ defined on $[-r,0]$, there is a unique solution x of (1.1) on $[-r,\infty)$. Designate this solution by $x(\varphi)$.

(II) The solution $x(\varphi)$ has a continuous derivative for $t > 0$, but not at $t = 0$ unless $\varphi(\theta)$ has a left hand derivative at $\theta = 0$ and $\dot{\varphi}(0) = -\varphi(-r)$. The solution $x(\varphi)$ is smoother than the initial data.

(III) For a given φ on $[-r,0]$, the solution $x(\varphi)(t)$ of (1.1) need not be defined for $t \leq -r$. In fact, if $x(\varphi)(t)$ is defined for $t \leq -r$, say $x(\varphi)(t)$ is defined for $t \geq -r-\varepsilon$, $\varepsilon > 0$, then $\varphi(\theta)$ must have a continuous first derivative for $\theta \in (-\varepsilon,0]$. If a solution $x(\varphi)$ does exist for $t \leq -r$, then $x(\varphi)(t)$ for $t \leq -r$ has in general fewer derivatives than φ.

Compare properties (II) and (III) with the parabolic partial differential equations.

As a second example, consider the <u>advanced</u> equation

$$\frac{dy(\tau)}{dt} = y(\tau+r), \quad r > 0. \tag{1.3}$$

If we let $\tau = -t$, $x(t) = y(-t)$, then x satisfies (1.1). Therefore, the natural problem for (1.3) is for $\tau \leq 0$. On the other hand, if this equation describes a physical system, then it must be integrated for $\tau \geq 0$. As in (III) above, any such solution must satisfy some special conditions and, in general, has fewer derivatives than the initial data.

As another example, consider the <u>neutral</u> equation

$$\dot{x}(t) - c\dot{x}(t-r) - dx(t-r) = 0, \quad r > 0,\ c \neq 0. \tag{1.4}$$

In this situation, it is a little more difficult to begin the discussion since many different possibilities are available for the concept of a solution.

In any case, if (1.4) is to define a function for $t \geq 0$, then we must specify a function on $[-r,0]$. If we suppose that φ is a function on $[-r,0]$ which has a continuous first derivative, then one can certainly find a function which satisfies (1.4) for $t > 0$ and even has a continuous first derivative except at the points $t = kr$, $k = 0,1,2,\ldots$. In fact $\dot{x}(t) = c\dot{x}(t-r) + dx(t-r)$ can be integrated successively in steps of length r. If $\dot{\varphi}(0) \neq c\dot{\varphi}(-r) + d\varphi(-r)$, then $\dot{x}(t)$ is discontinuous at $t = 0$. Consequently, $\dot{x}(t)$ will be discontinuous at $t = kr$, $k = 1,2,\ldots$. Since $c \neq 0$, we can also write

$$\dot{x}(t-r) = \frac{1}{c}[\dot{x}(t) - dx(t-r)]$$

and, therefore, define $x(t)$ for $t \leq -r$. The following observations are now immediate:

(IV) For any function φ defined on $[-r,0]$ with $\dot{\varphi}(\theta)$ continuous, there is a unique solution $x(\varphi)$ of (1.4) on $(-\infty,\infty)$ which has a continuous first derivative for $t \neq kr$, $k = 0, \pm 1, \pm 2,\ldots$.

(V) The solution $x(\varphi)$ has essentially the same smoothness properties as the initial data. Compare this with hyperbolic partial differential equations. One can also interpret (1.4) in integrated form as

$$x(t) - cx(t-r) = \varphi(0) - c\varphi(-r) + d\int_0^t x(s-r)ds, \quad t \geq 0.$$

A solution can now be defined for a continuous initial function. For $c = 0$, this now includes the retarded equation (1.2).

As a final example, consider the equation of <u>mixed type</u>

$$\dot{x}(t) + ax(t-r) + bx(t+r) = 0, \quad r > 0, \quad a \neq 0, \quad b \neq 0. \tag{1.5}$$

For this equation it is not at all clear what information is needed for (1.5) to define a function for $t \geq 0$ since the derivative of x depends upon past as well as future values. This equation seems to dictate that boundary conditions should be specified in order to obtain a solution in the same way as one does for elliptic partial differential equations.

Just looking at the examples above from the point of view of the information needed to obtain solutions of the equations and the resulting smoothness properties of the solutions, we have seen there are distinct types in a manner suggestive of the types in partial differential equations. To gain more insight into the differences in these types, let us look at their corresponding characteristic equations. As for linear ordinary differential equations with constant coefficients, the characteristic equation is obtained by trying to find a λ such that $e^{\lambda t}$ is a solution of the differential equation.

For equation (1.1), the characteristic equation is

$$\lambda + e^{-\lambda r} = 0. \tag{1.6}$$

It is clear that λ satisfies (1.6) if and only if

$$\lambda r + \ln \lambda = (2k+1)\pi i, \quad k = 0, \pm 1, \pm 2, \ldots$$

or

$$\operatorname{Re}(\lambda r + \ln r) = 0$$

$$r \operatorname{Re} \lambda = -\operatorname{Re} \ln |\lambda| .$$

Therefore,

$$\operatorname{Re} \lambda \to -\infty \quad \text{as} \quad |\lambda| \to \infty \quad \text{if} \quad r > 0 \quad \text{(retarded)}$$

$$\operatorname{Re} \lambda \to +\infty \quad \text{as} \quad |\lambda| \to \infty \quad \text{if} \quad r < 0 \quad \text{(advanced)}.$$

Since (1.6) is an entire function of λ, this implies there are only a <u>finite</u>

number of roots to the right of any line Re $z = \gamma$ if $r > 0$ (retarded) and there are only a finite number of roots to the left of any line Re $z = \gamma$ if $r < 0$ (advanced). Also, as $r \to 0^+$, Re $\lambda \to -\infty$ unless $|\lambda| \to 1$ and as $r \to 0^-$, Re $\lambda \to +\infty$ unless $|\lambda| \to 1$. It is natural to expect that the asymptotic behavior of the solutions will be depicted by the supremum of the real parts of the λ satisfying the characteristic equation. If this is so, then for $r \to 0^+$ the equation degenerates nicely (as far as asymptotic properties at $t = \infty$ are concerned) to the ordinary equation $\dot{x}(t) = -x(t)$.

A direct analysis of the characteristic equation of (1.4) is rather difficult. To contrast the difference with the equation of retarded type (1.1) and advanced type (1.3), it is convenient to begin with a rather degenerate situation. Consider the equation of neutral type

$$\dot{x}(t) - a\dot{x}(t-r) + bx(t) - abx(t-r) = 0, \quad a \neq 0, \quad r > 0. \tag{1.7}$$

The function $x(t) = e^{\lambda t}$ will be a solution of (1.7) if and only if λ is a solution of the characteristic equation

$$(\lambda+b)(1-ae^{-\lambda r}) = 0, \quad r > 0. \tag{1.8}$$

Therefore, $\lambda = -b$, $\lambda = (\ln a)/r + 2k\pi i/r$, $k = 0, \pm 1, \pm 2, \ldots$. Notice there are infinitely many roots to the right of some line Re $z = \gamma_1$ and infinitely many roots to the left of some line Re $z = \gamma_2$. Also, for $a \neq 1$, except for the root $\lambda = -b$, Re $\lambda \to \pm\infty$ as $r \to 0$.

Thus, the relationship of the equation (1.7) to the degenerate ordinary differential equation as $r \to 0$ is not even clear intuitively. On the other hand, if $|a| < 1$ in (1.8), then Re $\lambda \to -\infty$ as $r \to 0$ except for the root $\lambda = -b$. In this case, one can show that the solutions of (1.7) are asymptotic as $t \to \infty$ to the solutions of the degenerate ordinary differential equation as $r \to 0$.

The situation depicted above for (1.7) is typical for equations of neutral type. To be somewhat more convincing, consider the equation

$$\dot{x}(t) - a\dot{x}(t-r) + x(t) - x(t-r) = 0, \quad a \neq 0, \; r > 0. \tag{1.9}$$

The characteristic equation is

$$\lambda(1-ae^{-\lambda r}) + (1-e^{-\lambda r}) = 0 \tag{1.10}$$

or, for $\lambda \neq 0$,

$$e^{\lambda r}(1 + \frac{1}{\lambda}) - a - \frac{1}{\lambda} = 0.$$

Thus, if $|\lambda| \to \infty$, then λ must approach one of the roots of the equation $e^{\lambda r} = a$ or $\lambda r = \ln a + 2k\pi i$, $k = 0, \pm 1, \pm 2, \ldots$, that is, there are infinitely many roots λ in a vertical strip in the complex plane.

Due to the drastic differences in the behavior of solutions of the above simple examples, it is clear that one should attempt to classify the equations in some manner. For DDE, Bellman and Cooke [1] have given such a classification in terms of retarded, neutral and advanced type.

Our interest in these notes is on a geometric theory of FDE and not necessarily DDE. The above examples have indicated that some caution must be exercised even to isolate a class of FDE which will be small enough to have nice mathematical properties and yet large enough to include many interesting applications. Recently, a class of equations called neutral functional differential equations (NFDE) has been defined (see Hale and Cruz [1], Hale [8]) for which a geometric theory is rapidly evolving. This class is modeled after the observation that equation (1.4) can be represented in an integrated form to include the retarded equations. This class of NFDE also includes many of the current applications.

It certainly is tempting to present the general theory of NFDE. However, there are so many complications involved that it was decided to confine our attention to equations of retarded type. Whereever possible, proofs are given in such a way as to extend to neutral equations. Supplementary notes and references are given to permit the reader to delve further into NFDE if desired.

2. A GENERAL INITIAL VALUE PROBLEM

Suppose $r \geq 0$ is a given real number, $R = (-\infty,\infty)$, R^n is a real or complex n-dimensional linear vector space with norm $|\cdot|$, $C([a,b],R^n)$ is the Banach space of continuous functions mapping the interval $[a,b]$ into R^n with the topology of uniform convergence. If $[a,b] = [-r,0]$ we let $C = C([-r,0],R^n)$ and designate the norm of an element φ in C by $|\varphi| = \sup_{-r\leq\theta\leq 0}|\varphi(\theta)|$. Even though single bars are used for norms in different spaces, no confusion should arise. If $\sigma \in R$, $A \geq 0$ and $x \in C([\sigma-r,\sigma+A],R^n)$, then for any $t \in [\sigma,\sigma+A]$, we let $x_t \in C$ be defined by $x_t(\theta) = x(t+\theta)$, $-r \leq \theta \leq 0$. If $\cdot = d/dt$ and $f\colon R \times C \to R^n$ is a given function, we say that the relation

$$\dot{x}(t) = f(t,x_t) \tag{2.1}$$

is a functional differential equation of retarded type or simply a functional differential equation. A function x is said to be a <u>solution of</u> (2.1) if there are $\sigma \in R$, $A > 0$ such that $x \in C([\sigma-r,\sigma+A),R^n)$ and $x(t)$ satisfies (2.1) for $t \in (\sigma,\sigma + A)$. In such a case, we say x is a solution of (2.1) on $[\sigma-r,\sigma+A)$. For a given $\sigma \in R$ and a given $\varphi \in C$ we say $x = x(\sigma,\varphi)$ is a <u>solution of</u> (2.1) <u>with initial value</u> φ <u>at</u> σ or simply a <u>solution of</u> (2.1) <u>through</u> (σ,φ) if there is an $A > 0$ such that $x(\sigma,\varphi)$ is a solution of (2.1) on $[\sigma-r,\sigma+A)$ and $x_\sigma(\sigma,\varphi) = \varphi$.

Equation (2.1) is a very general type of equation and includes differential-difference equations

$$\dot{x}(t) = f(t,x(t),x(t-\tau(t)))$$

with $0 \leq \tau(t) \leq r$, as well as

$$\dot{x}(t) = \int_{-r}^{0} g(t,\theta,x(t+\theta))d\theta$$

and much more general types.

We say system (2.1) is linear if $f(t,\varphi) = L(t,\varphi) + h(t)$, where $L(t,\varphi)$ is linear in φ; linear, homogeneous if $h \equiv 0$ and linear nonhomogeneous if $h \not\equiv 0$. We say system (2.1) is autonomous if $f(t,\varphi) = g(\varphi)$ where g does not depend on t.

Lemma 2.1. If $\sigma \in R$, $\varphi \in C$ are given and $f(t,\varphi)$ is continuous, then finding a solution of (2.1) through (σ,φ) is equivalent to solving the integral equation

$$x(t) = \varphi(0) + \int_{\sigma}^{t} f(s,x_s)ds, \quad t \geq \sigma, x_\sigma = \varphi . \tag{2.2}$$

3. EXISTENCE

In this section, we give a basic existence theorem for the initial value problem of (2.1) assuming that f is continuous. We need

Lemma 3.1. If $x \in C([\sigma-r,\sigma+\alpha],R^n)$, then x_t is a continuous function of t for t in $[\sigma,\sigma+\alpha]$.

Proof. Since x is continuous on $[\sigma-r,\sigma+\alpha]$, it is uniformly continuous and thus for any $\varepsilon > 0$ there is a $\delta > 0$ such that $|x(t)-x(\tau)| < \varepsilon$ if $|t-\tau| < \delta$. Consequently, for t,τ in $[\sigma,\sigma+\alpha]$, $|t-\tau| < \delta$, we have $|x(t+\theta)-x(\tau+\theta)| < \varepsilon$ for all θ in $[-r,0]$. This proves the lemma.

Theorem 3.1. Suppose D is an open set in $R \times C$ and $f: D \to R^n$ is continuous. If $(\sigma,\varphi) \in D$, then there is a solution of (2.1) passing through (σ,φ).

Proof. For any real α,β, let

$$I_\alpha = \{t: 0 \leq t \leq \alpha\}, \quad B_\beta = \{\psi \in C: |\psi| \leq \beta\}.$$

If $|f(\sigma,\varphi)| < M$, then the continuity of f implies there are positive α,β such that $|f(\sigma+t,\varphi+\psi)| \leq M$ for $(t,\psi) \in I_\alpha \times B_\beta$. Choose α,β so that this latter relation is satisfied. For any nonnegative real $\bar{\alpha},\bar{\beta}$ let $\mathscr{A}(\bar{\alpha},\bar{\beta})$ be the set defined by

$$\mathscr{A}(\bar{\alpha},\bar{\beta}) = \{\eta \in C([-r,\bar{\alpha}],R^n): \eta_0 = 0, \ \eta_t \in B_{\bar{\beta}}, \ t \in I_{\bar{\alpha}}\}. \tag{3.1}$$

Suppose $\tilde{\varphi} \in C([\sigma-r,\sigma+\bar{\alpha}],R^n)$ is the function defined by $\tilde{\varphi}_\sigma = \varphi$, $\tilde{\varphi}(t+\sigma) = \varphi(0)$, $t \in I_{\bar{\alpha}}$. Suppose $\bar{\beta} < \beta$ and choose $\bar{\alpha}$ so that $|\tilde{\varphi}_{\sigma+t}-\varphi| < \beta-\bar{\beta}$, $t \in I_{\bar{\alpha}}$, $M\bar{\alpha} \leq \bar{\beta}$. Then $|\eta_t + \tilde{\varphi}_{t+\sigma}-\varphi| \leq \bar{\beta} + \beta - \bar{\beta} = \beta$ and $|f(\sigma+t,\eta_t+\tilde{\varphi}_{t+\sigma})| \leq M$ for $t \in I_{\bar{\alpha}}$, $\eta \in \mathscr{A}(\bar{\alpha},\bar{\beta})$. Consider the transformation $T:\mathscr{A}(\bar{\alpha},\bar{\beta}) \to C([-r,\bar{\alpha}],R^n)$ defined by

$$(3.2)\qquad \begin{aligned}(T\eta)(t) &= \int_0^t f(\sigma + s, \tilde{\varphi}_{\sigma+s} + \eta_s)ds, \quad t \in I_{\bar{\alpha}} \\ (T\eta)_0 &= 0\end{aligned}$$

Finding fixed points of T in $\mathscr{A}(\bar{\alpha},\bar{\beta})$ is equivalent to finding solutions of (2.1) in $\mathscr{A}(\bar{\alpha},\bar{\beta})$ since such fixed points η and solutions x of (2.1) through (σ,φ) are related by $x_{\sigma+t} = \tilde{\varphi}_{\sigma+t} + \eta_t$. We now apply the Schauder theorem to assert the existence of a fixed point of T in $\mathscr{A}(\bar{\alpha},\bar{\beta})$. The set $\mathscr{A}(\bar{\alpha},\bar{\beta})$ is a closed, bounded, convex subset of $C([-r,\bar{\alpha}],R^n)$.

Obviously, $T\eta, \eta$ in $\mathscr{A}(\bar{\alpha},\bar{\beta})$, is in $C([-r,\bar{\alpha}],R^n)$ and $(T\eta)_0 = 0$. Also,

$$|T\eta(t)| \le \int_\sigma^t |f(\sigma + s, \tilde{\varphi}_{\sigma+s} + \eta_s)|ds \le Mt \le M\bar{\alpha} \le \bar{\beta}$$

for $t \in I_{\bar{\alpha}}$. Therefore $|(T\eta)_t| \le \bar{\beta}$ for $t \in I_{\bar{\alpha}}$ and $T\mathscr{A}(\bar{\alpha},\bar{\beta}) \subset \mathscr{A}(\bar{\alpha},\bar{\beta})$. In addition

$$|T\eta(t) - T\eta(\bar{t})| \le |\int_{\bar{t}}^t f(\sigma + s, \tilde{\varphi}_{\sigma+s} + \eta_s)ds| \le M|t-\bar{t}|$$

for all $t,\bar{t}$ in $I_{\bar{\alpha}}$. Therefore $T\mathscr{A}(\bar{\alpha},\bar{\beta})$ belongs to a compact subset of $C([\sigma-r,\sigma+\bar{\alpha}],R^n)$.

It remains only to show that T is continuous on $\mathscr{A}(\bar{\alpha},\bar{\beta})$. Suppose η_k is a sequence in $\mathscr{A}(\bar{\alpha},\bar{\beta})$ which converges to η. Since $T\mathscr{A}(\bar{\alpha},\bar{\beta})$ belongs to a compact subset of $C([-r,\bar{\alpha}],R^n)$ there is a subsequence of the $\{\eta_k\}$ say $\{\eta_{k_j}\}$ such that $T\eta_{k_j} \to \gamma$ as $j \to \infty$. Since $f(\sigma+s,\tilde{\varphi}_{\sigma+s}+\eta_{k_j s}) \to f(\sigma+s,\varphi_{\sigma+s}+\eta_s)$ for all s in $I_{\bar{\alpha}}$ and f is bounded on $\mathscr{A}(\bar{\alpha},\bar{\beta})$, the Lebesgue dominated convergence theorem implies

$$\begin{aligned}\gamma(t) &= \lim_{j\to\infty}\left[\int_0^t f(\sigma+s,\tilde{\varphi}_{\sigma+s}+\eta_{k_j,s})ds\right] \\ &= \int_0^t f(\sigma+s,\tilde{\varphi}_{\sigma+s}+\eta_s)ds \\ &= (T\eta)(t), \qquad t \in I_{\bar{\alpha}}.\end{aligned}$$

This implies the limit of any convergent subsequence is independent of the subsequence. Since every subsequence of the $\{T\eta_k\}$ has a convergent subsequence and the limit $T\eta$ is independent of the subsequence, the sequence $T\eta_k$ itself converges to $T\eta$. Thus T is continuous on $\mathscr{A}(\bar{\alpha},\bar{\beta})$ and this completes the proof of the theorem.

4. CONTINUATION OF SOLUTIONS

Suppose f in (2.1) is continuous. If x is a solution of (2.1) on an interval $[\sigma,a)$, $a > \sigma$, we say $\hat{x}$ is a <u>continuation</u> of x if there is a $b > a$ such that $\hat{x}$ is defined on $[\sigma-r,b)$, coincides with x on $[\sigma-r,a)$, and $\hat{x}$ satisfies (2.1) on $[\sigma,b)$. A solution x is noncontinuable if no such continuation exists; that is, the interval $[\sigma,a)$ is the maximal interval of existence of the solution x. The existence of a noncontinuable solution follows from Zorn's lemma. Also, the maximal interval of existence must be open.

<u>Theorem 4.1.</u> Suppose D is an open set in $R \times C$ and $f: D \to R^n$ is continuous. If x is a noncontinuable solution of (2.1) on $[\sigma-r,b)$, then for any compact set W in D, there is a t_W such that $(t,x_t) \notin W$ for $t_W \le t < b$.

<u>Proof.</u> The case $b = \infty$ is trivial so we suppose b finite. Consider first the case $r = 0$ (an ordinary equation) and suppose the conclusion of the theorem is false. Then there are a sequence $t_k \to b$ as $k \to \infty$ and a $y \in R^n$ such that $(t_k,x(t_k)) \in W$, $(b,y) \in W$, $x(t_k) \to y$ as $k \to \infty$. There is a neighborhood U of (b,y) in D and a positive constant N such that $|f(t,x)| \le N$ for $(t,x) \in U$. Repeating the arguments in the basic existence theorem, there are an $\alpha > 0$ and a neighborhood V of (b,y) such that any solution of (2.1) for $r = 0$ through $(\tau,z) \in V$ exists on an interval $[\tau,\tau+\alpha)$. Consequently, for k sufficiently large, $(t_k,x(t_k)) \in V$ and the solution through this point exists on $[t_k,t_k+\alpha)$. Also, for k sufficiently large, $t_k + \alpha > b$. This contradicts the fact that x is noncontinuable on $[\sigma,b)$ and proves the theorem for $r = 0$.

If the conclusion of the theorem is not true for $r > 0$, then there are a sequence of real numbers $t_k \to b^-$ as $k \to \infty$ and a $\psi \in C$ such that $(t_k,x_{t_k}) \in W$, $(b,\psi) \in W$, $(t_k,x_{t_k}) \to (b,\psi)$ as $k \to \infty$. Thus, for any $\varepsilon > 0$,

$$\lim_{k \to \infty} \sup_{\theta \in [-r,-\varepsilon]} |x_{t_k}(\theta) - \psi(\theta)| = 0.$$

Since $x_t(\theta) = x(t+\theta)$, $-r \le \theta \le 0$ and $r > 0$, this implies $x(b+\theta) = \psi(\theta)$,

$-r \leq \theta < 0$. Hence $\lim_{t \to b^-} x(t)$ exists and x can be extended to a continuous function on $[\sigma-r,b]$ by defining $x(b) = \psi(0)$. Since $(b,x_b) \in D$, one can find a solution of (2.1) through this point to the right of b. This contradicts the non-continuability hypothesis on x and proves the theorem.

<u>Corollary 4.1.</u> Suppose D is an open set in $R \times C$ and $f: D \to R^n$ is continuous. If x is a noncontinuable solution of (2.1) on $[\sigma-r,b)$ and W is the closure of the set $\{(t,x_t), \sigma \leq t < b\}$ in $R \times C$, then W compact implies there is a sequence $\{t_k\}$ of real numbers, $t_k \to b^-$ as $k \to \infty$ such that (t_k,x_{t_k}) tends to ∂D as $k \to \infty$. If $r > 0$, then there is a $\psi \in C$ such that $(b,\psi) \in \partial D$ and $(t,x_t) \to (b,\psi)$ as $t \to b^-$.

<u>Proof.</u> Theorem 4.1 implies that W does not belong to D and proves the first part of the corollary. If $r > 0$, then the same argument as in the proof of Theorem 4.1 implies $\lim_{t \to b^-} x(t)$ exists and, thus, x can be extended as a continuous function on $[\sigma-r,b]$. Clearly, $(b,x_b) \in \partial D$ and $(t,x_t) \to (b,x_b)$ as $t \to b^-$.

<u>Theorem 4.2.</u> Suppose D is an open set in $R \times C$, $f: D \to R^n$ is continuous and takes closed bounded sets into bounded sets, and x is a noncontinuable solution of (2.1) on $[\sigma-r,b)$. Then for any closed bounded set U in D, there is a t_U such that $(t,x_t) \notin U$ for $t_U \leq t < b$.

<u>Proof.</u> The case $r = 0$ is contained in Theorem 4.1. Therefore, we suppose $r > 0$ and it is no restriction to take b finite. Suppose the conclusion of the theorem is not true. Then there is a sequence of real numbers $t_k \to b^-$ such that $(t_k,x_{t_k}) \in U$ for all k. Since $r > 0$, this implies that $x(t)$, $\sigma-r \leq t < b$ is bounded. Consequently, there is a constant M such that $|f(\tau,\varphi)| \leq M$ for (τ,φ) in the closure of $\{(t,x_t), \sigma \leq t < b\}$. The integral equation for the solutions of (2.1) imply

$$|x(t+\tau) - x(t)| = |\int_t^{t+\tau} f(s,x_s)ds| \leq M\tau$$

for all $t, t + \tau < b$. Thus, x is uniformly continuous on $[\sigma-r,b)$. This implies $\{(t,x_t), \sigma \leq t < b\}$ belongs to a compact set in D. This contradicts Theorem 4.1 and proves the theorem.

Theorem 4.2 shows that the trajectory (t,x_t) of a noncontinuable solution on $[\sigma,b)$ approaches the boundary of D as $t \to b$. The following example of Mishkis [1] shows that the behavior of a trajectory as $t \to b$, in general, can be very complicated and that the hypothesis in Theorem 4.2 that f take bounded sets into bounded sets cannot be eliminated.

Let $\Delta(t) = t^2$ and select two sequences $\{a_k\}$, $\{b_k\}$ of negative numbers, $a_1 < a_2 < \ldots$, $b_1 < b_2 < \ldots$, $a_k \to 0$, $b_k \to 0$ as $k \to \infty$ such that

$$a_k = b_k - \Delta(b_k), \quad b_k \leq a_{k+1} - \Delta(a_{k+1}), \quad k = 1,2,\ldots$$

For example, choose $b_k = -2^{-k}$, $k = 1,2,\ldots$.

Let $\psi(t)$ be an arbitrary continuously differentiable function satisfying

$$\begin{aligned} \psi(t) &= +1, \text{ for } t \text{ in } (-\infty,a_k], [b_{2k},a_{2k+1}], k = 1,2,\ldots \\ &= -1, \; t \in [b_{2k-1},a_{2k}], k = 1,2,\ldots \\ \psi'(t) &\neq 0, \; t \in [a_k,b_k], k = 1,2,\ldots \end{aligned}$$

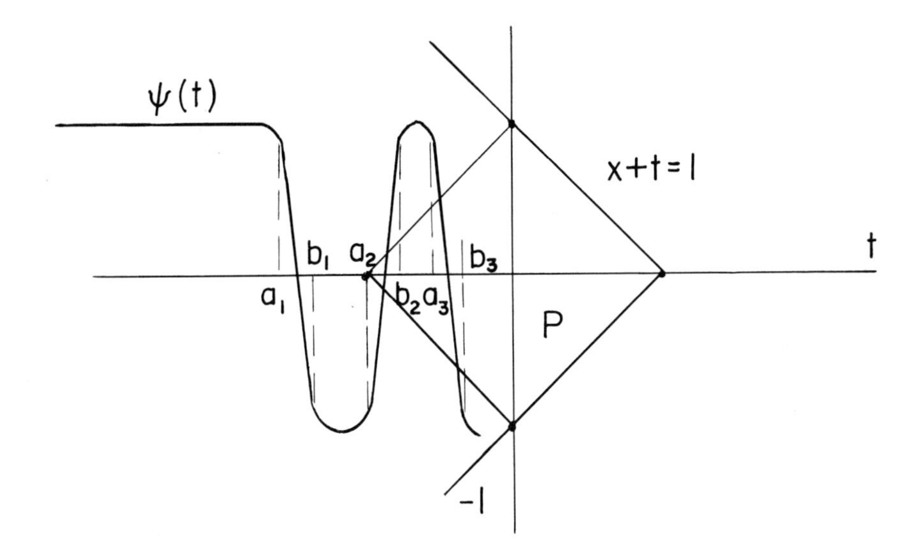

Let H be the set of points (t,x) such that $|x| < 1 - t$. These inequalities are equivalent to

$$x + t < 1 \quad \text{if} \quad x > 0$$
$$-x + t < 1 \quad \text{if} \quad x < 0.$$

Thus H is the wedge. We now define a function $h(t,x)$ on H. On the graph of the curve $\psi(t)$, let

$$h(t - \Delta(t),\ \psi(t - \Delta(t))) = \psi'(t), \quad ' = \text{derivative},\ -\infty < t < 0.$$

The function h is continuous on the graph of ψ. For any t in $[a_k,b_k]$, $k \geq 2$ (i.e., a point of increase or decrease on the graph), $t - \Delta(t) \in [b_{k-1},a_k]$. For t in $(-\infty,b_1]$, $t - \Delta(t) \in (-\infty,a_1]$. Therefore, $h = 0$ for any t in $(-\infty,b_1]$, $[a_k,b_k]$, $k \geq 2$, and, in particular, $h = 0$ on all points of increase or decrease of the graph of the curve $\psi(t)$. Now continue the function $h(t,x)$ in any manner whatsoever as long as it remains continuous and is equal to zero in the square $P: |t| + |x| \leq 1$. This can be done by linear interpolation with respect to x.

Now consider the equation

$$\dot{x}(t) = h(t - \Delta(t), y(t - \Delta(t))), \; t < 0, \; \Delta(t) = t^2. \tag{4.1}$$

Choose $\sigma < a_1$ and let $r = \sigma - \min\{(t-t^2), \sigma \leq t \leq 0\}$. We consider the initial value problem starting at σ. The function $y(t) = \psi(t)$ is a solution of this equation for $t < 0$ and is a non-continuable solution on $[\sigma-r,0)$.

If the right hand side of (4.1) is denoted by $f(t,x_t)$, $t \in R$, $x_t \in C([-r,0],R)$, then $f(t,\varphi)$ does not map closed bounded sets of $D = R \times C([-r,0],R)$ into bounded sets. In fact, the set $\{(t,x_t), t < 0\}$ is a bounded set and it is closed since there are no sequences $t_k \to 0$ such that ψ_{t_k} converges.

5. CONTINUOUS DEPENDENCE AND UNIQUENESS

Theorem 5.1. Suppose D is an open set in $R \times C$, $f_k\colon D \to R^n$, $k = 0,1,2,\ldots$ are continuous, $f_k(s,\psi) \to f_0(s,\varphi)$ as $k \to \infty$, $\psi \to \varphi$ for all $(s,\varphi) \in D$, and for every compact set W in D, there is an open neighborhood $V(W)$ of W and a constant $M > 0$ such that

$$(5.1) \qquad |f_k(t,\psi)| \le M, \quad (t,\psi) \in V(W), \quad k = 0,1,2,\ldots$$

Finally, suppose $\sigma_k \in R$, $\varphi_k \in C$, $(\sigma_k,\varphi_k) \in D$, $k = 0,1,2,\ldots$, $\sigma_k \to \sigma_0$, $\varphi_k \to \varphi_0$ as $k \to \infty$ and suppose $x^k = x^k(\sigma_k,\varphi_k)$, $k = 0,1,2,\ldots$ is a solution of

$$(5.2) \qquad \dot{x}^k(t) = f_k(t,x_t^k), \quad t \ge \sigma_k$$

with initial value φ_k at σ_k. If x^0 is defined on $[\sigma_0-r,b]$ and is the only solution through (σ_0,φ_0), then there is an integer k_0 such that each x^k, $k \ge k_0$, can be defined on $[\sigma_k-r,b]$ and $x^k(t) \to x^0(t)$ uniformly on $[\sigma_0-r,b]$. Since all x^k may not be defined on $[\sigma_0-r,b]$, by $x^k \to x^0$ uniformly on $[\sigma_0-r,b]$, we mean that for any $\varepsilon > 0$, there is a $k_0 = k_0(\varepsilon) \ge 0$ such that $x^k(t)$, $k \ge k_0$, is defined on $[\sigma-r-\varepsilon,b]$ and $x^k(t) \to x^0(t)$ uniformly for $t \in [\sigma-r-\varepsilon,b]$.

Proof. The set $W = \{(t,x_t^0), \sigma_0 \le t \le b\}$ is a compact subset of D. From the hypothesis on the f_k, there is an open neighborhood $V = V(W)$ of W and an $M > 0$, such that $|f_k(t,\psi)| \le M$, $(t,\psi) \in V$. Define $I_\alpha, B_\beta, \mathscr{A}(\alpha,\beta)$ as in the proof of Theorem 3.1 and let $\tilde{\varphi}^k \in C([\sigma_k-r,\sigma_k+\alpha),R^n)$ be defined by $\tilde{\varphi}^k_{\sigma_k} = \varphi_k$, $\tilde{\varphi}^k(\sigma_k+t) = \varphi_k(0)$, $t \in I_\alpha$. There is an open neighborhood V_1 of W, $V_1 \subset V$ and positive α,β so that $(t+\tau,\eta+\psi) \in V_1$ for any $(t,\eta) \in I_\alpha \times B_\beta, (\tau,\psi) \in V_1$. Suppose $\bar{\alpha},\bar{\beta}$ are chosen so that $2\bar{\alpha} < \alpha$, $2\bar{\beta} < \beta$, $M\bar{\alpha} < \beta$, $|\tilde{\varphi}^0_{t+\sigma}-\varphi_0| < \beta/2$ for $t \in I_{\bar{\alpha}}$. Since the φ_k, $k \ge 0$, form a compact set in C, the $\tilde{\varphi}^k$, $k \ge 0$, form a compact set in $C([\sigma_k-r,\sigma_k+\bar{\alpha}],R^n)$ and $\tilde{\varphi}^k \to \tilde{\varphi}^0$ as $k \to \infty$. Therefore, there is a

$k_0 \geq 0$ such that $|\sigma_k - \sigma_0| < \alpha/2$, $|\tilde{\varphi}^k_t - \varphi_k| < \beta/2$, $t \in I_{\bar{\alpha}}$, $k \geq k_0$. Thus, $|f_k(\sigma_k + t, \eta_t + \tilde{\varphi}^k_t)| \leq M$, for $t \in I_{\bar{\alpha}}$, $\eta \in \mathscr{A}(\bar{\alpha},\bar{\beta})$, $k = 0,1,2,\ldots$.

Now define the operators T_k, $k = 0,1,2,\ldots$, taking $\mathscr{A}(\bar{\alpha},\bar{\beta})$ into $C([0,\bar{\alpha}],R^n)$ by the relations

$$(T_k\eta)(t) = \int_0^t f_k(\sigma_k + s, \tilde{\varphi}^k_{\sigma_k+s} + \eta_s)ds, \quad t \in I_{\bar{\alpha}}$$

$$(T_k\eta)_0 = 0.$$

Since $\tilde{\varphi}^k \to \tilde{\varphi}^0$ as $k \to \infty$, $f_k(t,\psi) \to f_0(t,\varphi)$ as $k \to \infty$, $\psi \to \varphi$ and the f_k are uniformly bounded on V, it follows from the Lebesgue dominated convergence theorem that $T_k\eta \to T_0\eta$ for each $\eta \in \mathscr{A}(\bar{\alpha},\bar{\beta})$. Each of the operators T_k is a compact mapping of $\mathscr{A}(\bar{\alpha},\bar{\beta})$ into $\mathscr{A}(\bar{\alpha},\bar{\beta})$. Thus, from the Schauder theorem, there is a fixed point $\eta^k \in \mathscr{A}(\bar{\alpha},\bar{\beta})$ of T_k. On the other hand, the family of functions $\eta^k = T_k\eta^k \in \mathscr{A}(\bar{\alpha},\bar{\beta})$ are equicontinuous and uniformly bounded. Therefore, there is a subsequence which we again index with k and $\eta^k \to \eta^*$ as $k \to \infty$. Since $T_k\eta \to T_0\eta$ for every $\eta \in \mathscr{A}(\bar{\alpha},\bar{\beta})$ it follows that $\eta^* = \eta^0$. Since every convergent subsequence of the η^k must converge to η^0, it follows that $\eta^k \to \eta^0$ as $k \to \infty$. Due to the compactness of the set $\{(t,x^0_t), t \in [\sigma^0,b]\}$, one completes the proof by successively stepping intervals of length $\bar{\alpha}$.

<u>Theorem 5.2.</u> Suppose D is an open set in $R \times C$, $f\colon D \to R^n$ is continuous and $f(t,\varphi)$ is Lipschitzian in φ in each compact set in D. If $(\sigma,\varphi) \in D$, then there is a unique solution of (2.1) with initial value φ at σ.

<u>Proof.</u> Define I_α, B_β as in the proof of Theorem 3.1 and suppose x,y are solutions of (2.1) on $[\sigma-r, \sigma+\alpha]$ with $x_\sigma = \varphi = y_\sigma$. Then

$$x(t) - y(t) = \int_\sigma^t [f(s,x_s) - f(s,y_s)]ds, \quad t \geq \sigma.$$

$$x_\sigma - y_\sigma = 0.$$

If k is the Lipschitz constant of $f(t,\varphi)$ in any compact set containing the trajectories (t,x_t), (t,y_t), $t \in I_\alpha$, then choose $\bar{\alpha}$ so that $k\bar{\alpha} < 1$. Then, for $t \in I_{\bar{\alpha}}$,

$$|x(t) - y(t)| \leq \int_\sigma^t k|x_s - y_s|\,ds$$

$$\leq k\bar{\alpha} \sup_{\sigma \leq s \leq t} |x_s - y_s|$$

and this implies $x(t) = y(t)$ for $t \in I_{\bar{\alpha}}$. One completes the proof of the theorem in an obvious manner.

Theorems on the differentiability of the solutions of (2.1) with respect to parameters as well as Frechét differentiability with respect to initial data are supplied in exactly the same manner as in ordinary differential equations. It is convenient to obtain the differentiability of $x(\sigma,\varphi)$ in σ,φ by first proving differentiability with respect to φ and then using the observation that

$$x_t(\sigma,\varphi) = x_t(\sigma + h, x(\sigma + h, x_{\sigma+h}(\sigma,\varphi))).$$

6. BACKWARD CONTINUATION

We say a function $x \in C([\sigma-r-\alpha,\sigma],R^n)$, $\alpha > 0$, is a <u>solution of (2.1) on $[\sigma-r-\alpha,\sigma]$ through (σ,φ)</u> if $x_\sigma = \varphi$ and for any $\sigma_1 \in [\sigma-r-\alpha,\sigma]$, x is a solution of (2.1) on $[\sigma_1-r,\sigma]$ through (σ_1,x_{σ_1}).

<u>Definition 6.1.</u> If D is an open set in $R \times C$, we say a function $f: D \to R^n$ is <u>atomic at $-r$ on D</u> if $f(t,\varphi)$ has a continuous Frechét derivative $f'_\varphi(t,\varphi)$ with respect to φ on D, and there exist an $n \times n$ matrix $\eta(t,\varphi,\theta)$ of bounded variation in θ for $(t,\varphi) \in D$, $\theta \in [-r,0]$, an $n \times n$ matrix $A(t,\varphi)$, $\det A(t,\varphi) \neq 0$ and a scalar function $\gamma(t,\varphi,s)$ continuous for $(t,\varphi) \in D$, $s \geq 0$, $\gamma(t,\varphi,0) = 0$ such that

$$f'_\varphi(t,\varphi)\psi = \int_{-r}^{0} [d_\theta\eta(t,\varphi,\theta)]\psi(\theta),$$

$$\eta(t,\varphi,-r^+) - \eta(t,\varphi,-r) = A(t,\varphi)$$

$$\left|\int_{-r}^{-r+s} [d_\theta\eta(t,\varphi,\theta)]\psi(\theta) - A(t,\varphi)\psi(-r)\right| \leq \gamma(t,\varphi,s) \sup_{-r\leq\theta\leq-r+s} |\psi(\theta)|.$$

<u>Theorem 6.1.</u> If D is an open set in $R \times C$, $f: D \to R^n$ is atomic at $-r$ on D, $(\sigma,\varphi) \in D$ and there is an $0 < \alpha < r$ such that $\dot\varphi(\theta)$ is continuous for $\theta \in [-\alpha,0]$, $\dot\varphi(0) = f(\varphi)$, then there is an $\bar\alpha > 0$ and a unique solution x of (2.1) on $[\sigma-r-\bar\alpha,\sigma]$ through (σ,φ).

<u>Proof.</u> A function x is a solution of (2.1) on $[\sigma-r-\alpha,\sigma]$ through (σ,φ) if and only if $x_\sigma = \varphi$, $(t,x_t) \in D$, $t \in [\sigma-\alpha,\sigma]$, and

$$f(t,x_t) = \dot x(t), \quad t \in [\sigma-\alpha,\sigma]. \tag{6.1}$$

For any $\alpha > 0$, let $\tilde\varphi: [\sigma-r-\alpha,\sigma] \to R^n$, $\tilde\varphi(t) = \varphi(t)$, $t \in [\sigma-r,\sigma]$, $\tilde\varphi(t) = \varphi(-r)$, $t \in [\sigma-r-\alpha,\sigma-r]$. Then x is a solution of (6.1) if and only if $x = \tilde\varphi + y$ and y satisfies

$$f(t,\tilde{\varphi}_t+\tilde{y}_t) = \dot{\varphi}(t-\sigma) \quad t \in [\sigma-\alpha,\sigma]$$

$$y_\sigma = 0.$$

If we let $z(t) = y(\sigma+t)$, $\hat{\varphi}(t) = \tilde{\varphi}(\sigma+t)$, $t \in [\alpha,0]$, then x is a solution of (6.1) if and only if $x(\sigma+t) = \hat{\varphi}(t) + z(t)$ and $z(t)$ satisfies

$$f(\sigma+t,\hat{\varphi}_t+z_t) = \dot{\varphi}(t), \quad t \in [-\alpha,0]. \tag{6.2}$$

If $f(t,\varphi+\psi) = f(t,\varphi) + f'_\varphi(t,\varphi)\psi + g(t,\varphi,\psi)$, then the definition of the derivative implies that $g(t,\varphi,\psi)$ is continuous in (t,φ,ψ), $g(t,\varphi,0) = 0$ and

$$|g(t,\varphi,\psi) - g(t,\varphi,\xi)| \le \varepsilon(t,\varphi,\beta)|\psi-\xi|, \ |\psi|, \ |\xi| \le \beta$$

where $\varepsilon(t,\varphi,\beta)$ is continuous in (t,φ,β) for $(t,\varphi) \in D$, $\beta \ge 0$, and $\varepsilon(t,\varphi,0) = 0$. If we make use of this in (6.2), then x is a solution of (6.1) if and only if $x(\sigma+t) = \hat{\varphi}(t) + z(t)$ and $z(t)$ satisfies

$$f'_\varphi(\sigma+t,\hat{\varphi}_t)z_t = -f(\sigma+t,\hat{\varphi}_t) - g(\sigma+t,\hat{\varphi}_t,z_t) + \dot{\varphi}(t), \ t \in [-\alpha,0], \tag{6.3}$$

$$z_0 = 0.$$

But using the definition of $A(t,\varphi)$ we see that $x(\sigma+t) = \hat{\varphi}(t) + z(t)$ is a solution of (6.1) if and only if $z(t)$ satisfies

$$\begin{aligned} z(t) &= A^{-1}(\sigma+t,\varphi_t)\{\int_{-r^+}^{0}[d_\theta\eta(t,\hat{\varphi}_t,\theta)]z_t(\theta) - f(\sigma+t,\hat{\varphi}_t) \\ &\quad -g(\sigma+t,\hat{\varphi}_t,z_t) + \dot{\varphi}(t)\}, \quad t \in [-\alpha,0] \\ z_0 &= 0 \end{aligned} \tag{6.4}$$

For any $\beta > 0$, let $B_\beta = \{\psi \in C: |\psi| \le \beta\}$. For any ν, $0 < \nu < 1/4$, there are $\alpha > 0$, $\beta > 0$, such that $(\sigma+t,\varphi+\psi) \in D$,

$$|A^{-1}(\sigma+t,\varphi+\psi)|\,\varepsilon(\sigma+t,\varphi+\psi,\beta) < \nu$$

$$|A^{-1}(\sigma+t,\varphi+\psi)|\,\gamma(\sigma+t,\varphi+\psi,\alpha) < \nu .$$

Choose α,β so that these relations are satisfied. For any non-negative real $\bar{\alpha},\bar{\beta}$, let $\mathscr{A}(\bar{\alpha},\bar{\beta})$ be the set defined by

$$\mathscr{A}(\bar{\alpha},\bar{\beta}) = \{\zeta \in C([-r-\bar{\alpha},0],R^n):\ \zeta_0 = 0,\ \zeta_t \in B_{\bar{\beta}},\ t \in [-\bar{\alpha},0]\}.$$

For any $0 < \bar{\beta} < \beta$, there is an $\bar{\alpha}$, $0 < \bar{\alpha} < \alpha$, so that $|\hat{\varphi}_t-\varphi| < \beta - \bar{\beta}$. Thus $|\zeta_t+\hat{\varphi}_t-\varphi| \leq \bar{\beta} + \beta - \bar{\beta} = \beta$ and $(\sigma+t,\hat{\varphi}_t+\xi_t) \in D$ for $t \in [-\bar{\alpha},0]$, $\zeta \in \mathscr{A}(\bar{\alpha},\bar{\beta})$. Further restrict $\bar{\alpha}$ so that

$$|f(\sigma+t,\hat{\varphi}_t) - f(\sigma,\varphi)| \leq \nu\bar{\beta}$$

$$|\dot{\varphi}(0) - \dot{\varphi}(t)| \leq \nu\bar{\beta}$$

for $t \in [-\bar{\alpha},0]$.

For any $\zeta \in \mathscr{A}(\bar{\alpha},\bar{\beta})$, define the transformation $T: \mathscr{A}(\bar{\alpha},\bar{\beta}) \to C([-r-\bar{\alpha},0], R^n)$ by the relation

$$(T\zeta)(t) = 0, \qquad t \in [-r,0]$$

$$(6.5) \qquad (T\zeta)(t) = A^{-1}(\sigma+t,\hat{\varphi}_t)\{\int_{-r^+}^{0} [d_\theta\eta(t,\hat{\varphi}_t,\theta)]\zeta_t(\theta) - f(\sigma+t,\hat{\varphi}_t) + f(\sigma,\varphi) - g(\sigma+t,\hat{\varphi}_t,\zeta_t) + \dot{\varphi}(t) - \dot{\varphi}(0)\},\quad t \in [-\bar{\alpha},0].$$

By hypothesis $\dot{\varphi}(0) = f(\sigma,\varphi)$ and therefore the fixed points of T in $\mathscr{A}(\bar{\alpha},\bar{\beta})$ coincide with the solutions x of (6.2) on $[\sigma-r-\alpha,\sigma]$ with $x(\sigma+t) = \hat{\varphi}(t) + \zeta(t)$ where $\zeta_t \in \mathscr{A}(\bar{\alpha},\bar{\beta})$, $t \in [-\bar{\alpha},0]$.

We now show that T is a contraction on $\mathscr{A}(\bar{\alpha},\bar{\beta})$. It is clear from (6.5) and the above restrictions on $\bar{\alpha},\bar{\beta}$ that

$$|(T\zeta)(t)| \leq \nu\bar{\beta} + \nu\bar{\beta} + \nu\bar{\beta} + \nu\bar{\beta} = 4\nu\bar{\beta} \leq \bar{\beta}$$

$$|(T\zeta)(t) - (T\xi)(t)| \leq \nu|\zeta_t - \xi_t| + \nu|\zeta_t - \xi_t| \leq \frac{1}{2}|\zeta_t - \xi_t|$$

for all $t \in [-\alpha,0]$, $\zeta, \xi \in \mathscr{A}(\bar{\alpha},\bar{\beta})$. Therefore, $T: \mathscr{A}(\bar{\alpha},\bar{\beta}) \to \mathscr{A}(\bar{\alpha},\bar{\beta})$ and is a contraction. Thus, there is a unique fixed point in $\mathscr{A}(\bar{\alpha},\bar{\beta})$ and this proves the theorem.

Corollary 6.1. Suppose D is an open set in $R \times C$, $f: D \to R^n$ is continuous, atomic at $-r$ on D, and the solution $x(\sigma,\varphi)$ of (2.1) through any (σ,φ) for $(\sigma,\varphi) \in D$ is unique. If $T(t,\sigma): C \to C$, $t \geq \sigma$, is defined by $T(t,\sigma)\varphi = x_t(\sigma,\varphi)$, then $T(t,\sigma)$ is one-to-one.

Proof. If the assertion is not true, then there are $\psi \neq \varphi$ in C and a $t_1 > \sigma$ such that $x_{t_1}(\sigma,\varphi) = x_{t_1}(\sigma,\psi)$, $x_t(\sigma,\varphi) \neq x_t(\sigma,\psi)$, $\sigma \leq t < t_1$. If $x(t) = x(\sigma,\varphi)(t)$, $y(t) = x(\sigma,\psi)(t)$, then $\dot{x}(t) = f(t,x_t)$, $\dot{y}(t) = f(t,y_t)$ for all $t > 0$ in the domain of definition of x. Since f is assumed to be atomic at $-r$, Theorem 6.1 implies there are an $\alpha = \alpha(t_1) > 0$ and a unique solution of (2.1) on $[t_1-r-\alpha, t_1-r]$ through (t_1,x_{t_1}), (t_1,y_{t_1}). Since $(t_1,x_{t_1}) = (t_1,y_{t_1})$ by hypotheses, it follows that $(t,x_t) = (t,y_t)$ for $t_1 - \alpha \leq t \leq t_1$. This is a contradiction and proves the corollary.

As a first example, consider the linear system

$$\dot{x}(t) = \int_{-r}^{0} [d_\theta \eta(t,\theta)]x(t+\theta) \stackrel{\text{def}}{=} L(t,x_t) \tag{6.6}$$

where $\eta(t,-r^+) - \eta(t,-r) = A(t)$ is continuous and

$$\left|\int_{-r}^{-r+s} [d_\theta \eta(t,\theta)]\psi(\theta) - A(t)\psi(-r)\right| \leq \gamma(t,s) \sup_{-r\leq\theta\leq-r+s} |\psi(\theta)| \tag{6.7}$$

for a continuous scalar function $\gamma(t,s)$, $t \in R$, $s \geq 0$, $\gamma(t,0) = 0$. If $\det A(t) \neq 0$ for all t, then $L(t,\varphi)$ is atomic at $-r$ on $R \times C$ and the map $T(t,\sigma)$ defined by the solutions of (6.6) is one-to-one.

Even if $\det A(t) = 0$ for some t, the following relation is true. For any $\delta > 0$, one can find an $n \times n$ matrix $B(t)$ such that $B(t)$ is continuous for $t \in R$, $|B(t)| < \delta$ for $t \in R$ and $\det[A(t) + B(t)] \neq 0$ for $t \in R$. Therefore, the map defined by the solutions of the equation

$$\dot{x}(t) = L(t,x_t) + B(t)x(t-r) \overset{\text{def}}{=} f(t,x_t)$$

is one-to-one since $f(t,\varphi)$ is atomic at $-r$ on $R \times C$ and $|f(t,\varphi) - L(t,\varphi)| \leq \delta|\varphi|$. Consequently, by an arbitrarily small change in the linear operator $L(t,\cdot)$, one can be assured that the mapping is one-to-one.

As another example, consider the equation

$$\dot{x}(t) = L(t,x_t) + N(t,x_t) \overset{\text{def}}{=} F(t,x_t) \tag{6.8}$$

where L is the same function as in (6.6), $N(t,\varphi)$ is continuous for $(t,\varphi) \in R \times C$ and the Frechét derivative $N_\varphi(t,\varphi)$ of N with respect to φ is continuous and $|N_\varphi(t,\varphi)| \leq \mu(|\varphi|)$ for $(t,\varphi) \in R \times C$, where μ is a continuous function with $\mu(0) = 0$. If $|\det A(t)| \geq a > 0$ for $t \in R$, then $F(t,\varphi)$ is atomic at $-r$ on $R \times U$ where U is a sufficiently small neighborhood of the origin in C. Consequently, the map $T(t,\sigma)$ defined by the solutions of (6.8) is one-to-one on its domain of definition.

As a final example, consider the equation

$$\dot{x}(t) = -\alpha x(t-1)[1+x(t)], \quad \alpha > 0. \tag{6.9}$$

For this case, $f(t,\varphi) = -\alpha\varphi(-1)[1+\varphi(0)]$ and

$$f'_\varphi(t,\varphi)\psi = -\alpha\psi(-1)[1+\varphi(0)] - \alpha\varphi(-1)\psi(0)$$

$$A(t,\varphi) = -\alpha[1+\varphi(0)].$$

As long as $\varphi(0) \neq -1$, the function $f(t,\varphi)$ is atomic at $-r$ and the mapping $T(t,\sigma)$ will be one-to-one as long as the solution $x(\sigma,\varphi)(t) \neq -1$. But, from the equation,

$$x(t) = -1 + [1+\varphi(0)]e^{-\int_\sigma^t \alpha x(s-1)ds}, \quad t \geq \sigma.$$

Therefore, any solution with $\varphi(0) \neq -1$ will always have $x(t) \neq -1$ and $T(t,\sigma)$ defined by $T(t,\sigma)\varphi = x_t(\sigma,\varphi)$ is one-to-one on the sets $\{\varphi \in C: \varphi(0) > -1\}$ and $\{\varphi \in C: \varphi(0) < -1\}$. On the set, $C_{-1} \overset{def}{=} \{\varphi \in C: \varphi(0) = -1\}$, the map $T(t,\sigma)$ is not one-to-one and, in fact, $T(t,\sigma)\varphi$ is the constant function one for $t \geq \sigma + 1$ and $\varphi \in C_{-1}$.

It is natural to ask whether one can approximate the right-hand side $f(t,\varphi)$ of (6.9) by a function $g(t,\varphi)$ which is atomic at $-r$ on $R \times C$ and which is close to $f(t,\varphi)$ on closed bounded sets. The map induced by the new equation would be one-to-one. It is easy to see that such a g does not exist. In fact, if

$$g'_\varphi(t,\varphi) = \int_{-r}^{0} [d_\theta \nu(t,\varphi,\theta)]\psi(\theta)$$

$$\nu(t,\varphi,-r^+) - \nu(t,\varphi,-r) = B(t,\varphi)$$

then $B(t,\varphi)$ must be close to $-\alpha[1+\varphi(0)]$ on closed, bounded sets. This implies $B(t,\varphi)$ must have a zero for some φ. This problem of approximation needs to be discussed in much more detail.

7. CARATHEODORY CONDITIONS

In Section 2, we defined a functional differential equation for continuous $f: R \times C \to R^n$. On the other hand, it was then shown that the initial value problem was equivalent to

$$(7.1) \qquad \begin{aligned} x_\sigma &= \varphi \\ x(t) &= \varphi(0) + \int_\sigma^t f(s,x_s)ds, \quad t \geq \sigma. \end{aligned}$$

The equation is certainly meaningful for a more general class of functions f if it is not required that $x(t)$ has a continuous first derivative for $t > \sigma$. We give in this section the appropriate generalization to functional differential equations of the well known Caratheodory conditions of ordinary differential equations.

Suppose D is an open subset of $R \times C$. A function $f: D \to R^n$ is said to satisfy the Caratheodory condition on D if $f(t,\varphi)$ is measurable in t for each fixed φ, continuous in φ for each fixed t and for any fixed $(t,\varphi) \in D$, there is a neighborhood $V(t,\varphi)$ and a Lebesgue integrable function m such that

$$(7.2) \qquad |f(s,\psi)| \leq m(s), \quad (s,\psi) \in V(t,\varphi).$$

If $f: D \to R^n$ is continuous, it is easy to see that f satisfies the Caratheodory condition on D. Therefore, a theory for (7.1) in this more general setting will include the previous theory.

If f satisfies the Caratheodory conditions on D, $(\sigma,\varphi) \in D$, we say a function $x = x(\sigma,\varphi)$ is a solution of (7.1) through (σ,φ) if there is an $A > 0$ such that $x \in C([\sigma-r,\sigma+A],R^n)$, $x_\sigma = \varphi$, $x(t)$ is absolutely continuous on $[\sigma,\sigma+A]$ and satisfies $\dot{x}(t) = f(t,x_t)$ almost everywhere on $[\sigma,\sigma+A]$.

Using essentially the same arguments, one can extend all of the previous results to the case where f satisfies a Caratheodory condition on D. Of course, in the analogue of Theorem 5.1, all f_k should satisfy the Caratheodory condition on D, $f_k(s,\psi) \to f_0(s,\varphi)$ as $k \to \infty$, $\psi \to \varphi$ for almost all s and condition (5.1) should be replaced by the following: For any compact set W in U, there is an open neighborhood V(W) of W and a Lebesgue integrable function M such that the sequence of functions f_k, $k = 0,1,2,\ldots$, satisfy

$$|f_k(s,\psi)| \leq M(s), \quad (s,\psi) \in V(W)$$
$$k = 0,1,2,\ldots .$$

We remark in passing that more general existence theorems are easily given if the function $f(t,\varphi)$ depends upon φ in some special way. In particular, if for any $\varepsilon \geq 0$ we let φ^{ε} denote the restriction of φ to the interval $[-r,-\varepsilon]$ and

$$f(t,\varphi) = F(t,\varphi(0),\varphi^{\varepsilon}),$$

then the basic existence theorem can be proved by the process of stepping forward a step of size less than ε (if $\varepsilon > 0$) under very weak conditions on the dependence of $F(t,x,\psi)$ upon ψ.

8. REMARKS ON THE MAP DEFINED BY SOLUTIONS

In this section, we give some specific examples of functional differential equations in order to contrast the behavior with ordinary differential equations. The examples will also serve to familiarize the reader with the idea of looking at the solution of (2.1) in the space C rather than R^n.

Throughout this section, suppose $f: R \times C \to R^n$ is continuous and for any $(\sigma,\varphi) \in R \times C$, there is a unique solution $x = x(\sigma,\varphi)$ of (2.1) passing through (σ,φ). From Theorem 6.1, $x(\sigma,\varphi)(t)$ is continuous in (t,σ,φ) in its domain of definition.

<u>Remark 8.1.</u> <u>Two distinct solutions of</u> (2.1) <u>considered in</u> $R \times R^n$ <u>may intersect an infinite number of times.</u> In fact, consider the scalar equation

$$\dot{x}(t) = -x(t-\pi/2)$$

which has the solutions $x = \sin t$, $x = \cos t$. The sets $\{(t,\sin t),\ t \in R\}$, $\{(t, \cos t),\ t \in R\}$ in $R \times R$ intersect an infinite number of times.

The above example shows that it is probably impossible to develop a geometric theory for (2.1) by defining trajectories in $R \times R^n$ as $\{(t, x(\sigma,\varphi)(t)),\ t \geq \sigma\}$.

On the other hand, it seems reasonable to have the definition of a trajectory of a solution so that it will depict the evolution of the state of the system. Furthermore, the state of the system should be that part of the system which uniquely determines the future behavior. From our basic existence and uniqueness theorem the state at time t therefore, should be $x_t(\sigma,\varphi)$ and the <u>trajectory through</u> (σ,φ) should be the set $\bigcup_{t \geq 0} (t,x_t(\sigma,\varphi))$ in $R \times C$. For the geometric theory of functional differential equations, the map is going to be $x_t(\sigma,\cdot)$. Therefore, for $t \geq \sigma$, define the operator $T(t,\sigma): C \to C$ by the relation

$$(8.1) \qquad T(t,\sigma)\varphi = x_t(\sigma,\varphi).$$

The operator $T(t,\sigma)$ is continuous. From the hypothesis of uniqueness of solutions of (2.1), for given $\varphi, \psi \in C$, if there is a $\tau \geq \sigma$ such that $T(\tau,\sigma)\varphi = T(\tau,\sigma)\psi$, then $T(t,\sigma)\varphi = T(t,\sigma)\psi$ for $t \geq \tau$.

Even if $\psi \neq \varphi$, the possibility is not excluded that the two trajectories through (σ,ψ), (σ,φ) may intersect at some $\tau > \sigma$. In fact, an example where this occurs was given by equation (6.9). To reemphasize this remark and to discuss some more geometry of the solutions, we give another example where $T(t,\sigma)$ is not one-to-one.

<u>Remark 8.2.</u> <u>The operator</u> $T(t,\sigma)$ <u>need not be one-to-one.</u> Consider the example

$$(8.2) \qquad \dot{x}(t) = -\alpha x(t-1)[1-x^2(t)].$$

Equation (8.2) has the solution $x(t) = 1$ for all t in $(-\infty,\infty)$. Furthermore, if $r = 1$, $\sigma = 0$ and $\varphi \in C$, then there is a unique solution $x(0,\varphi)$ of (8.2) through $(0,\varphi)$ which depends continuously upon φ and if $-1 \leq \varphi(0) \leq 1$, these solutions are actually defined on $[-1,\infty)$. On the other hand, if $\varphi \in C$, $\varphi(0) = 1$, then $x(0,\varphi)(t) = 1$ for all $t \geq 0$. Therefore, for all such initial values, $x_t(0,\varphi)$, $t \geq 1$, is the constant function 1. A translate of a subspace of C of codimension one is mapped into a point by $T(t,0)$ for all $t \geq 1$.

The function $x(t) = -1$ is also a solution of (8.2) and for any $\varphi \in C$, $\varphi(0) = -1$, the solution $x(0,\varphi)(t)$ is -1 for $t \geq 0$. Therefore, $x_t(0,\varphi)$ is the constant function -1 for $t \geq 1$.

For this example, it is interesting to try to depict the trajectories in $R \times C$. For any constant a let $C_a = \{\varphi \in C: \varphi(0) = a\}$. The set C_a is the translate of a subspace of C of codimension 1 (a hyperplane) and $R \times C$ can be represented schematically as in the accompanying diagram. We have put on this diagram

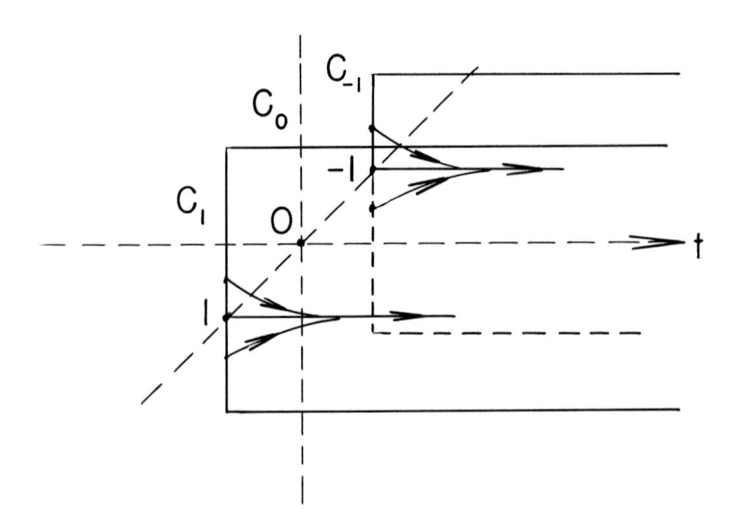

the sets $R \times C_1$ and $R \times C_{-1}$ as well as the constant functions 1 and -1 and representative trajectories in these planes. Notice that solutions are trapped between these planes if the initial values φ satisfy $-1 \leq \varphi(0) \leq 1$. Also notice that any solution which oscillates about zero must have a trajectory which crosses the set $R \times C_0$.

The fact that the map $T(t,\sigma)$ need not be one-to-one is very disturbing. Sufficient conditions for one-to-oneness were given in Theorem 6.1, but it is instructive to look at the general situation in a little more detail. Suppose $\Omega = R \times C$ and all solutions $x(\sigma,\varphi)$ of (2.1) are defined on $[\sigma-r,\infty)$. We say $(\sigma,\varphi) \in R \times C$ is equivalent to $(\sigma,\psi) \in R \times C$, $(\sigma,\varphi) \sim (\sigma,\psi)$ if there is a $\tau \geq \sigma$ such that $x_\tau(\sigma,\varphi) = x_\tau(\sigma,\psi)$; that is (σ,φ) is equivalent to (σ,ψ) if the trajectories through (σ,φ) and (σ,ψ) have a point in common. It is easy to see that "$\sim$" is an equivalence relation and the space C is decomposed into equivalence classes $\{V_\alpha\}$ for each fixed σ. If $T(t,\sigma)$ is one-to-one, then each equivalence class consists of a single point; namely, the initial value (σ,φ). For each equivalence class V_α choose a representative element $\varphi^{\sigma,\alpha}$ and let

$$W(\sigma) = \bigcup_\alpha \varphi^{\sigma,\alpha}. \tag{8.3}$$

From the point of view of the qualitative theory of functional differential equations, the set $W(\sigma)$ is very interesting since it is a maximal set on which the map $T(t,\sigma)$ is one-to-one. However, it seems to be very difficult to say much about the properties of $W(\sigma)$. In fact, without some more precise description of the manner in which $\varphi^{\sigma,\alpha}$ is chosen from V_α, one cannot hope to discuss such topological properties of $W(\sigma)$ as connectedness. For example, consider the scalar equation

$$\dot{x}(t) = 0$$

considered as a functional differential equation with lag $r > 0$. If $C_a = \{\varphi \in C: \varphi(0) = a\}$, then $\varphi \in C_a$ implies $x_t(\sigma,\varphi)$ is the constant function a for $t \geq \sigma + r$. Therefore, the equivalence classes V_α are the sets C_α, $-\infty < \alpha < \infty$, for each σ. An arbitrary choice of $\varphi^{\sigma,\alpha}$ leads to a very uninteresting set $W(\sigma)$. On the other hand, $W(\sigma)$ consisting of all the constant functions is certainly the set that is of interest for the equation. In a general situation, we know nothing about the "appropriate" choice of $\varphi^{\sigma,\alpha}$. The following examples are given to indicate some of the other difficulties involved.

Remark 8.3. _For autonomous linear equations,_ $W(0)$ _is completely determined in a finite time interval and can be chosen as a linear subspace of_ C. In fact, for an autonomous linear equation, D. Henry [1] has shown there is a number τ such that if $x_t(0,\varphi) = x_t(0,\psi)$ for $t \geq t_0$, and t_0 is chosen as small as possible, then $t_0 \leq \tau$; that is, the equivalence classes V_α are completely determined in the interval $[0,\tau]$. Let $T(t,0) = T(t)$ and consider the set $S = \{\varphi \in C: T(t)\varphi = 0,\ t \geq \tau\}$. This is a closed linear subspace of C invariant under $T(t)$. The set S admits projection in C (continuous ?), $C = S \oplus U$ where U is also invariant under $T(t)$. Furthermore, $T(t)$ is one-to-one on U. Thus, we can take $W(0) = U$ and each element of U corresponds to one of the equivalence classes V_α.

Remark 8.4. _For autonomous linear equations, the time interval for the determina-_

tion of $W(0)$ is in general $> r$; that is $T(t,0)$ need not be one-to-one on $T(r,0)C$. The following example was given by D. Henry. Suppose

$$A = \begin{bmatrix} 0 & 1 & 0 \\ 0 & 0 & 1 \\ 0 & 0 & 0 \end{bmatrix}$$

and $\dot{x}(t) = Ax(t-1)$, $t \geq 0$. If $x = (x_1,x_2,x_3)$ and $\varphi \in C([-1,0],R)$ is arbitrary, this system has the solution

$$x_3(t) = \begin{cases} \sin \pi t, & -1 \leq t \leq 0 \\ 0 \ , & t \geq 0 \end{cases}$$

$$x_2(t) = \begin{cases} \frac{2}{\pi} & , \ -1 \leq t \leq 0 \\ \frac{1}{\pi} + \frac{1}{\pi}\cos \pi t & , \ 0 \leq t \leq 1, \\ 0 & , \ t \geq 1 \end{cases}$$

$$x_1(t) = \begin{cases} \varphi(t) & , \ -1 \leq t \leq 0 \\ \frac{2}{\pi}t - \frac{3}{\pi} & , \ 0 \leq t \leq 1 \\ \frac{t-2}{\pi} - \frac{1}{\pi^2} \sin \pi t & , \ 1 \leq t \leq 2 \\ 0 & , \ t \geq 2 \end{cases}$$

The lag is 1 and the solution does not become zero until after $t \geq 3$.

Remark 8.5. For nonlinear equations, the equivalence classes V_α may involve the consideration of trajectories which have a point in common after any preassigned times. The following example was given by A. Hausrath. For $\beta > 0$, $r = 1$, consider the scalar equation

$$\dot{x}(t) = \beta[|x_t| - x(t)].$$

For a given φ in $C = C([-1,0],R)$, there is a unique solution $x = x(\varphi,\beta)(t)$ of this equation through $(0,\varphi)$ which is continuous in (φ,β,t).

If $\varphi(0) \geq 0$, $\varphi \neq 0$, then $x(\varphi,\beta)(t)$ is a positive constant for $t \geq 1$. In fact, since $\dot{x}(t) \geq 0$, it follows that $|x_t| = x(t)$ for $t \geq 1$ and uniqueness implies $x(t)$ is a constant $\geq \varphi(0)$ for $t \geq 1$. Also, if $\varphi(0) = 0$, then $\varphi \neq 0 \Rightarrow \dot{x}(0) > 0$ and $x(t) > 0$ for $t \geq 1$. Therefore, for any positive constant function, the corresponding equivalence class contains more than one element. Also, the above argument and the autonomous nature of the equation show that the equivalence class corresponding to the constant function zero contains only zero.

If $\varphi(0) < 0$, then it is clear that $x(\varphi,\beta)(t)$ approaches a constant as $t \to \infty$. If $x(\varphi,\beta)(t)$, $\varphi(0) < 0$, has a zero $z(\varphi,\beta)$, it must be simple and, therefore, $z(\varphi,\beta)$ is continuous in φ,β. For any $\beta > 0$, there exists a $\varphi \in C$ such that $z(\varphi,\beta)$ exists. In fact, let $\varphi \in C$, $\varphi(0) = -1$, $\varphi(\theta) = -\gamma$, $\gamma > 1$, $-1 \leq \theta \leq -1/2$ and let $\varphi(\theta)$ be a monotone increasing function for $-1/2 \leq \theta \leq 0$. As long as $x(t) \leq 0$ and $0 \leq t \leq 1/2$, we have $|x_t| = \gamma$ and

$$\dot{x}(t) = \beta[\gamma - x(t)] \geq \beta\gamma.$$

Therefore, $x(t) \geq \beta\gamma t - 1$. For $\beta\gamma/2 > 1$, it follows that x must have a zero.

The closed subset $C_{-1} = \{\varphi \in C\colon \varphi(0) = -1\}$ can be written as $C_{-1} = C^o_{-1} \cup C^n_{-1}$ where $C^o_{-1} = \{\varphi \in C_{-1}\colon z(\varphi,\beta)$ exists$\}$, $C^n_{-1} = \{\varphi \in C_{-1}\colon z(\varphi,\beta)$ does not exist$\}$. Since $z(\varphi,\beta)$ is continuous, the set C^o_{-1} is open and, therefore, C^n_{-1} is closed. For any $\varphi \in C^n_{-1}$, $x(\varphi,\beta)(t) \to 0$ monotonically as $t \to \infty$. Therefore, if C^n_{-1} is not empty, then there is a sequence $\varphi_j \in C^o_{-1}$, $\varphi_j \to \varphi \in C^n_{-1}$ as $j \to \infty$ and $z(\varphi_j,\beta) \to \infty$ as $j \to \infty$.

There is a $\beta_0 > 0$ such that C^n_{-1} is not empty. In fact, choose $\beta_0 > 0$ less than or equal to that value of β for which the equation $\lambda + \beta = -\beta e^{-\lambda}$ has a real root λ_0 of multiplicity two. For this β_0, the equation $\lambda + \beta = -\beta e^{-\lambda}$ has two real negative roots. If λ_0 is one of these roots, then $x(t) = -e^{-\lambda_0 t}$ is a solution of the above equation with initial value $\varphi_0(\theta) = -e^{-\lambda_0\theta}$, $-r \leq \theta \leq 0$,

$\varphi_0 \in C_{-1}$. Therefore, C_{-1}^n is not empty.

With β_0 as above, it follows that $\delta(\beta_0) \overset{\text{def}}{=} \sup_{\varphi \in C_{-1}^0} z(\varphi,\beta_0) = \infty$. Therefore, using the fact that our original equation is positive homogeneous of degree 1 in x, it follows that for any positive constants a, t_0, there exists a $\varphi \in C$, such $x(\varphi,\beta_0)(t) = a$, $t \geq t_0$, $x(\varphi,\beta_0)(t) < a$ for $0 \leq t \leq t_0$. This proves the assertion in the remark.

In D. Henry [1], it is shown that linear autonomous equations have the property that no two distinct solutions can exist on $(-\infty,\infty)$ and coincide on $[0,\infty)$. The following remark asserts this statement is false for nonlinear equations.

Remark 8.6. <u>There may be two distinct solutions of a FDE defined on</u> $(-\infty,\infty)$ <u>and yet they coincide on</u> $[0,\infty)$. The following example was given by A. Hausrath. Let $r = 1$, $f(s) = 0$, $0 \leq s \leq 1$, $f(s) = -3(\sqrt[3]{s} - 1)^2$, $s > 1$, and consider the equation

$$\dot{x}(t) = f(|x_t|).$$

The function $x \equiv 0$ is a solution of this equation on $(-\infty,\infty)$. Also, the function $x(t) = -t^3$, $t < 0$, $= 0$, $t \geq 0$ is also a solution. In fact, since $x \leq 1$ for $t \geq -1$, it is clear that x satisfies the equation for $t \geq 0$. Since x is monotone decreasing for $t \leq 0$, $|x_t| = x(t-1) = -(t-1)^3$ and $\dot{x}(t) = -3t^2$. It is easy to verify that $-3t^2 = f((1-t)^3)$ for $t < 0$.

Remark 8.7. <u>The map</u> $T(t,\sigma)$ <u>is locally bounded for any</u> $t \geq \sigma$; that is, for any $t \geq \sigma$, $\varphi \in C$, there is a neighborhood $V(t,\sigma,\varphi)$ of φ such that $T(t,\sigma)V(t,\sigma,\varphi)$ is bounded. This is an immediate consequence of the continuity of $T(t,\sigma)\varphi$ in φ. The fact that $T(t,\sigma)\varphi$ is continuous in t,σ,φ actually implies the following stronger result: For any $T > 0$, $\sigma \in R$, $\varphi \in C$, $\varepsilon > 0$, there is a neighborhood $V(\varepsilon,\sigma,\varphi,T)$ of φ such that

$$|T(t,\sigma)\psi - T(t,\sigma)\varphi| < \varepsilon, \quad \sigma \leq t \leq \sigma+T, \quad \psi \in V(\varepsilon,\sigma,\varphi,T).$$

Remark 8.8. <u>$T(t,\sigma)$ may not take closed bounded sets of C into bounded sets of C.</u> The following example is due to K. Hannsgen. Suppose $r = 1/4$, $C = C([-r,0],R)$ and consider the equation

$$\dot{x}(t) = f(t,x_t) \overset{\text{def}}{=} x^2(t) - \int_{\min\{t-r,0\}}^{0} |x(s)|\,ds. \tag{8.4}$$

It is clear that f takes closed bounded sets into bounded sets and is locally Lipschitzian. If $B = \{\varphi \in C: |\varphi| \leq 1\}$ and $x(b)$, $b \in B$, is the solution of (8.4), then $x(b)$ is always ≥ -1. Also, $\dot{x}(b)(0^+) < 1$ for all $b \in B$ and, thus, there is a $\sigma > 0$, independent of b such that $x(b)(\sigma) < (1-\sigma)^{-1}$. If $y(t,\sigma,x(b)(\sigma))$, $y(\sigma,\sigma,x(b)(\sigma)) = x(b)(\sigma)$ is the solution of $\dot{y}(t) = y^2(t)$, then $-1 \leq x(b)(t) \leq y(t,\sigma,x(b)(\sigma)) < (1-t)^{-1}$, $\sigma \leq t \leq r$. Thus, $x(b)(t)$ exists for $-r \leq t \leq r$ and $x(b)(r) < (1-r)^{-1}$ for all $b \in B$. For $t \geq r$, $\dot{x}(b)(t) = x^2(b)(t)$ and the fact that $x(b)(r) < (1-r)^{-1}$ implies $x(b)(t)$ exists for $-r \leq t \leq 1$.

If we show that for any $\varepsilon > 0$, there is a $b \in B$ such that $x(b)(r) > (1-r)^{-1} - \varepsilon$, then the set $x(B)(1)$ is not bounded. To show this, suppose $\varepsilon > 0$ is given, $C = |1-r|^{-1}$, $M = 2Cre^{2Cr} + 1$. Choose $b \in B$ so that $b(0) = 1$, $\int_{t-r}^{0} |b(t)|\,dt < \varepsilon/M$ and let $y(t) = y(t,0,1)$, $y(0,0,1) = 1$, be the solution of $\dot{y}(t) = y^2(t)$ and $x(t) = x(b)(t)$. If $\psi(t) = y(t) - x(t)$ for $0 < t < r$, then $\psi(t) \geq 0$ and $\dot{\psi}(t) \leq 2C\psi(t) + \varepsilon/M$. Since $\psi(0) = 0$, one thus obtains $\psi(r) \leq \varepsilon$. This shows that $x(r) = y(r) - \psi(r) = (1-r)^{-1} - \psi(r) \geq (1-r)^{-1} - \varepsilon$ and proves the general assertion made above.

Remark 8.9. <u>For any $t \geq \sigma + r$, the map $T(t,\sigma)$ is locally completely continuous</u>; that is, for any $t \geq \sigma + r$, $\varphi \in C$, there is a neighborhood $V(t,\sigma,\varphi)$ of φ such that $T(t,\sigma)V(t,\sigma,\varphi)$ is in a compact subset of C. In fact, Remark 8.7 implies $T(t,\sigma)$ is a locally bounded map. Since f is also assumed continuous, for any $\tau \geq r$, there is a neighborhood $V(\tau,\sigma,\varphi)$ of φ such that

$|T(t,\sigma)V(\tau,\sigma,\varphi)| \leq M$, $|f(t,T(t,\sigma)V(\tau,\sigma,\varphi))| \leq N$ for some constants M,N and $\sigma \leq t \leq \sigma + \tau$. Thus $|\dot{x}(\sigma,V(\tau,\sigma,\varphi))(t)| \leq N$ for $\sigma \leq t \leq \sigma + \tau$. For any $t \in [\sigma+r,\sigma+T]$, this implies the family of functions $\{x_t(\sigma,\psi), \psi \in V(T,\sigma,\varphi)\}$ belongs to a compact subset of C and proves the remark.

Remark 8.10. *Suppose $T(t,\sigma)$ takes closed bounded sets of C into bounded sets of C and $T(t,\sigma)$ is defined for all $t \geq \sigma$. If f takes closed bounded sets of $R \times C$ into bounded sets of R^n and U is any closed bounded subset of C, then $T(t,\sigma)U$ belongs to a compact subset of C for each $t \geq \sigma + r$.* In fact, fix an interval $[\sigma,\sigma+T]$, $T \geq r$. From the hypothesis, there is a constant $M > 0$ such that $|x_t(\sigma,\varphi)| \leq M$, $t \in [\sigma,\sigma+T]$, $\varphi \in U$. Since f maps closed bounded sets of $R \times C$ into bounded sets of R^n, there is a constant $N > 0$ such that $|\dot{x}(\sigma,\varphi)(t)| \leq N$, for $t \in [\sigma,\sigma+T]$, $\varphi \in U$. For any fixed $t \in [\sigma+r, \sigma+T]$, this implies the family of functions $\{x_t(\sigma,\varphi), \varphi \in U\}$ belongs to a compact subset of C and proves the result.

Remark 8.11. *Even if $T(t,\sigma)$ is one-to-one, it cannot be a homeomorphism for $r > 0$ and $t \geq \sigma + r$.* In fact, if $T(t,\sigma)$ were a homeomorphism, it would map closed bounded sets of C into bounded sets of C. But Remark 8.10 implies that any closed bounded set U is mapped by $T(t,\sigma)$ into a compact set. If $r > 0$, U is not compact, which proves the result.

Remark 8.12. *There are functional differential equations such that there is a $t_0 > 0$ with $[T(t_0,\sigma)C](0)$ zero dimensional.* In particular, R^n is not contained in this set, a phenomenon which is not possible in ordinary differential equations. Consider the equation

$$\dot{x}(t) = b(t)x(t - 3\pi/2) \tag{8.5}$$

where for an arbitrary $g \in C([-\infty,0],R^n)$

$$(8.6) \qquad b(t) = \begin{cases} g(t) & t \le 0 \\ 0 & 0 \le t \le 3\pi/2 \\ -\cos t & 3\pi/2 \le t \le 3\pi \\ 1 & t \ge 3\pi \end{cases} .$$

For $\sigma = 0$ and any arbitrary $\varphi \in C([-3\pi/2,0],R)$, the equation (8.5) has the solution

$$(8.7) \qquad x(0,\varphi)(t) = \begin{cases} \varphi(0) & , \quad 0 \le t \le 3\pi/2 \\ (-\sin t)\varphi(0) & , \quad t \ge 3\pi/2 \end{cases} .$$

Thus, $x(0,\varphi)(t) = 0$ for $t = k\pi$, $k = 2,3,\ldots$ and all $\varphi \in C([-3\pi/2,0],R)$.

An even more striking example is the following.

Remark 8.13. <u>There are functional differential equations for which there is a $t_0 > 0$ with $T(t,\sigma)C = \{0\}$ for all $t \ge t_0$.</u> Consider the equation

$$\dot{x}(t) = -\alpha(t)x(t-1)$$

where

$$\alpha(t) = 2\sin^2 \pi t, \quad t \in [2n,2n+1]$$
$$= 0 \quad , \quad t \in (2n-1,2n)$$

for each integer n. For any $\sigma \in R$, $\varphi \in C$, we show $T(t,\sigma)\varphi = 0$, $t \ge \sigma + 4$. In fact, if N is the smallest odd integer such that $N \ge \sigma$, then $x(t) = x(N)$, $t \in [N,N+1]$ and

$$\dot{x}(t) = -\alpha(t)x(N), \quad t \in [N+1,N+2].$$

Thus,

$$x(N+2) = x(N)[1-2\int_{N+1}^{N+2} \sin^2 \pi s ds] = 0.$$

Therefore, $x(t) = 0$ for $t \in [N+2,N+3]$ and $x(t) = 0$ for $t \geq N+2$.

9. AUTONOMOUS SYSTEMS

Due to the remarks of Section 8, it is impossible to develop a theory of autonomous functional differential equations which is as comprehensive as the one for ordinary differential equations. However, some of the meaningful concepts are given in this section.

Suppose $f: C \to R^n$ is continuous, takes closed bounded sets of C into bounded sets of R^n. If $x(\varphi)$ is the solution of the autonomous equation

$$\dot{x}(t) = f(x_t) \tag{9.1}$$

through $(0,\varphi)$, we also suppose that the solution is defined on $[-r,\infty)$, is unique, and therefore $x(\varphi)(t)$ is continuous in t,φ by Theorem 6.1. Any solution $x(\varphi)$ of (9.1) satisfies

$$x_0(\varphi) = \varphi$$

$$x_{t+\tau}(\varphi) = x_t(x_\tau(\varphi)), \quad t,\tau \geq 0$$

and, therefore, defines a dynamical system. The <u>orbit</u> $\gamma^+(\varphi)$ <u>through</u> φ is $\gamma^+(\varphi) = \bigcup_{t\geq 0} x_t(\varphi)$. An <u>equilibrium</u> (or <u>critical</u>) <u>point of</u> (9.1) is a constant function c such that $f(c) = 0$. A <u>periodic solution of</u> (9.1) is a solution p of (9.1) such that $p_{t+\omega} = p_t$ for some constant ω. It is easy to verify that a <u>nonconstant solution</u> $x(\varphi)$ <u>of</u> (9.1) <u>is periodic if and only if the orbit through</u> φ <u>is a closed curve</u>. A function $g(t)$, $-\infty < t < \infty$, is said to be a <u>solution of</u> (9.1) on $(-\infty,\infty)$ if for every $\sigma \in (-\infty,\infty)$, the solution $x(g_\sigma)$ of (9.1) satisfies $x_t(g_\sigma) = g_{t+\sigma}$, $t \geq 0$.

An element $\psi \in C$ is said to be in the ω-<u>limit set</u> $\omega(\varphi)$ of an orbit $\gamma^+(\varphi)$ through φ if there is a sequence of $t_n \to \infty$ as $n \to \infty$ such that $x_{t_n}(\varphi) \to \psi$ as $n \to \infty$. An element $\psi \in C$ is said to be in the α-<u>limit set</u> $\alpha(\varphi)$ of an orbit through φ if there is a solution $x(\varphi)$ of (9.1) on $(-\infty,\infty)$, $x_0(\varphi) = \varphi$, and a sequence $t_n \to -\infty$ as $n \to \infty$ such that $x_{t_n}(\varphi) \to \psi$ as $n \to \infty$.

The α- and ω-limit sets can also be characterized by

$$\alpha(\varphi) = \bigcap_{\tau \leq 0} \overline{\bigcup_{t \leq \tau} x_t(\varphi)}$$

$$\omega(\varphi) = \bigcap_{\tau \geq 0} \overline{\bigcup_{t \geq \tau} x_t(\varphi)} .$$

A set M in C is said to be <u>invariant</u> if for every $\varphi \in M$, there is a solution x of (9.1) on $(-\infty,\infty)$ with $x_0 = \varphi$, $x_t \in M$, $t \in (-\infty,\infty)$.

<u>Lemma 9.1.</u> If there is a constant $m > 0$ and a solution x of (9.1) with $|x(t)| < m$, $t \in [\sigma-r,\infty)$, then $\gamma^+(x_0)$ belongs to a compact subset of C.

<u>Proof.</u> Since $|x_t| < m$, $t \geq 0$ and f maps closed bounded sets into bounded sets, there is a constant $N > 0$ such that $|\dot{x}(t)| \leq N$, $t \geq 0$. Ascoli's theorem yields the result.

Uniqueness of the solutions of (9.1) is not used in the following lemma.

<u>Lemma 9.2.</u> If there is an M and a solution $x(t)$ of (9.1) on $[-r,\infty)$ such that $|x(t)| \leq M$ for $t \geq -r$, then $\omega(\gamma^+(x_0))$ is a nonempty, compact connected invariant set and

$$\text{dist}(x_t,\omega(\gamma^+(x_0))) \to 0 \quad \text{as} \quad t \to \infty.$$

<u>Proof.</u> The above Lemma 9.1 implies $\gamma^+(x_0)$ belongs to a compact subset of C. Therefore, $\omega(\gamma^+(x_0))$ is nonempty and bounded. To prove $\omega(\gamma^+(x_0))$ is closed, suppose $\psi_n \in \omega(\gamma^+(x_0))$, $\psi_n \to \psi \in C$. There are sequences $x_{t_{k,n}} \to \psi_n$ as $k \to \infty$, $t_{k,n} \to \infty$ as $k \to \infty$. For any $\varepsilon > 0$, there is a $k(\varepsilon)$, $n(\varepsilon)$ such that $x_{t_{k(\varepsilon),n(\varepsilon)}}$ is in the ε-neighborhood of ψ. Choose a sequence of $\varepsilon_j \to 0$ as $j \to \infty$ to obtain the closure of $\omega(\gamma^+(x_0))$. As a consequence, $\omega(\gamma^+(x_0))$ is nonempty and compact.

It is obvious that $\text{dist}(x_t,\omega(\gamma^+(x_0))) \to 0$ as $t \to \infty$. For otherwise, there would be an $\varepsilon > 0$ and a sequence $t_k \to \infty$ as $k \to \infty$ such that $\mathrm{dist}(x_{t_k}, \omega(\gamma^+(x_0))) > \varepsilon$, $k = 1,2,\ldots$. The sequence x_{t_k} belongs to a compact subset of C and, therefore, must have a convergent subsequence. The limit point must be in $\omega(\gamma^+(x_0))$, which is a contradiction.

This latter result obviously implies $\omega(\gamma^+(x_0))$ is connected. Otherwise, $\omega(\gamma^+(x_0))$ would be the union of two disjoint compact sets which are a distance δ apart. Since $x_t \to \omega(\gamma^+(x_0))$ as $t \to \infty$, this would imply a sequence of $t_k \to \infty$ as $k \to \infty$ such that $\mathrm{dist}(x_{t_k}, \omega(\gamma^+(x_0))) > \delta/2$, which is a contradiction.

It remains only to show invariance. Suppose $\psi \in \omega(\gamma^+(x_0))$ and let $t_k \to \infty$ monotonically as $k \to \infty$ be such that $x_{t_k} \to \psi$ as $k \to \infty$. For any integer $N \geq 0$, there is a $k_0(N)$ such that $x(t+t_k)$ is defined for $-N - r \leq t \leq N$, $k \geq k_0(N)$. The sequence $\{x(t+t_k)\}$ is uniformly bounded and equicontinuous on $-N-r \leq t \leq N$. Consequently, there is a subsequence $\{x(t+t_{k,N})\}$ and a continuous function $g(t)$ defined on $[-N-r,N]$ with $x(t+t_{k,N}) - g(t) \to 0$ as $k \to \infty$ uniformly on $[-N-r,N]$. Using the diagonalization procedure, we can find a subsequence of the t_k which we label the same as before and a continous function $g(t)$, $-\infty < t < \infty$, such that the sequence $x(t+t_k) - g(t) \to 0$ as $k \to \infty$ uniformly on compact subsets of $(-\infty,\infty)$. (Note that no elements of the sequence may be defined on $(-\infty,\infty)$, but the assertion remains valid all the same.) Also, $x_{t+t_k} \to g_t$ as $k \to \infty$ uniformly on compact subsets of $(-\infty,\infty)$. Therefore $g_t \in \omega(\gamma^+(x_0))$, $t \in (-\infty,\infty)$.

Next, we show that $g(t)$ is continuously differentiable and $\dot{g}(t) = f(g_t)$. In fact, for t in any compact set $[-N,N]$, there is a $K(N)$ such that

$$\begin{aligned} &|g(t+h) - g(t) - hf(g_t)| \\ &\quad \leq |g(t+h) - x(t+h+t_k)| + |x(t+h+t_k) \\ &\qquad\quad - x(t+t_k) - hf(x_{t+t_k})| + |x(t+t_k) - g(t)| \\ &\qquad\quad + h|f(x_{t+t_k}) - f(g_t)| \end{aligned}$$

for all $k \geq K(N)$. Choose $k(h)$ in such a way that $k(h) \to \infty$ as $h \to 0$ and

$|g(t) - x(t+t_{k(h)})| = o(|h|)$ as $h \to 0$ for all t in $[-N,N]$. The right hand side of this inequality is now $o(|h|)$ as $h \to 0$ which proves $\dot{g}(t) = f(g_t)$. Therefore, g is a solution of (1) on $(-\infty,\infty)$ and $g_t \in \omega(\gamma^+(x_0))$, $t \in (-\infty,\infty)$, $g_0 = \psi$. This proves invariance and the lemma.

10. DEFINITIONS OF STABILITY

Let $C_H = \{\varphi \in C: |\varphi| < H\}$, $R^+ = [0,\infty)$. In this section, we consider the system (3.1) with $f(t,0) \equiv 0$, $t \in R^+$, $f: R^+ \times C_H$ is continuous and satisfies enough additional hypotheses to ensure that the solution $x(\sigma,\varphi)(t)$ through (σ,φ) is continuous in (σ,φ,t) in the domain of definition of the function.

Definition 10.1. (a) The solution $x = 0$ of (3.1) is called stable at t_0 if $t_0 \geq 0$ and

(i) there is a $b = b(t_0) > 0$ such that φ in C_b implies the solution $x(t_0,\varphi)$ of (3.1) exists for $t \geq t_0$ and $x_t(t_0,\varphi)$ is in C_H for $t \geq t_0$;

(ii) For every $\varepsilon > 0$, there is a $\delta = \delta(t_0,\varepsilon) > 0$ such that φ in C_δ implies the solution $x(t_0,\varphi)$ of (3.1) satisfies $x_t(t_0,\varphi)$ in C_ε for all $t \geq t_0$.

(b) The solution $x = 0$ of (3.1) is called asymptotically stable at t_0 if it is stable and there is an $H_0 = H_0(t_0)$ such that φ in C_{H_0} implies the solution $x(t_0,\varphi)$ of (3.1) satisfies

$$\lim_{t\to\infty} |x_t(t_0,\varphi)| = 0.$$

(c) The solution $x = 0$ of (3.1) is unstable at t_0 if it is not stable at t_0.

In ordinary differential equations, a system which enjoys either one of the above types of stability at t_0 enjoys the same type of stability at t_1 for any $t_1 \geq t_0$. The basic reason for this fact is that the mapping induced by the solutions of ordinary differential equations for which solutions are uniquely defined by their initial values takes a sphere of initial values into a set which contains a sphere. Also, continuity with respect to initial values implies the above remark is also true for any $t_1 \leq t_0$ provided only that solutions of the equation exist on $[t_1,t_0]$.

For functional-differential equations, the latter property holds for exactly the same reason; namely, <u>if the solution $x = 0$ of (1.1) is stable at t_0 in the sense of definition (10.1) (a) or (10.1) (b), then it is stable at $t_1 \le t_0$ in the same sense provided that the solutions exist on $[t_1, t_0]$.</u>

However, stability of the solution $x = 0$ of (3.1) at t_0 does not necessarily imply stability of $x = 0$ at $t_1 > t_0$. In fact, consider equation (8.5). For $t_0 = 0$, the solution is given by (8.7) and so the solution $x = 0$ of (8.5) is clearly stable for $t_0 = 0$. On the other hand, for any $t_1 > 3\pi$, the solution $x(t,\varphi)$ of (8.5) must satisfy the equation

$$\dot{x}(t) = x(t - 3\pi/2). \tag{10.1}$$

For any constant a and any λ for which $\lambda = \exp(-3\pi\lambda/2)$, the function $x(t) = a \exp \lambda t$ satisfies (10.1). Since there is a $\lambda_0 > 0$ satisfying this equation, the solution $x = 0$ is unstable for any $t_1 > 3\pi$.

It seems to be very difficult to determine in general when stability at t_0 implies stability at $t_1 \ge t_0$, but the following result is very easy and indicates the simplicity of autonomous and periodic systems.

<u>Lemma 10.1.</u> If $f(t,\varphi)$ is either independent of t or periodic in t, then stability (or asymptotic stability) of $x = 0$ of (3.1) at t_0 implies stability (or asymptotic stability) of $x = 0$ of (3.1) at all $t_1 \ge t_0$.

<u>Proof.</u> If the period of f is T then stability at t_0 implies stability at $t_0 + kT$ for any positive integer k. But, stability at $t_0 + kT$ implies stability at all $t_1 \le t_0 + kT$ for which the solutions are defined on $[t_1, t_0+kT]$. Continuous dependence of solutions on initial values implies there is a $b = b(kT)$ such that for any φ in C_b and t_1 in $[t_0, t_0+kT]$, the solution $x(t_1,\varphi)$ is defined on $[t_1, t_0+kT]$. The lemma is proved.

For a scalar equation, the following result is true.

<u>Lemma 10.2.</u> For x a scalar, consider the equation

$$(10.2) \qquad \dot{x}(t) = \int_{-r}^{0} x(t+\theta)d\eta(t,\theta) \overset{\text{def}}{=} L(t,x_t)$$

where $L(t,\varphi)$ is continuous for $(t,\varphi) \in R^+ \times C$. If $\eta(t,\theta)$ is nondecreasing in θ for each fixed t, then stability of the solution $x = 0$ at t_0 implies stability for all $t \geq t_0$.

Proof. It is sufficient to prove that the solution $x = 0$ of (10.2) is stable at t_0 if and only if $\int_{t_0}^{\infty} [\eta(t,0) - \eta(t,-r)]dt$ exists. If $x = 0$ is stable at t_0, and $\varphi \in C$ is defined as $\varphi(\theta) = c$, $0 < c < \delta$ for all θ, then $x(t) = x(t_0,\varphi)(t)$ is nondecreasing,

$$\dot{x}(t) \geq c \int_{-r}^{0} d\eta(t,\theta) = c[\eta(t,0) - \eta(t,-r)]$$

and, thus,

$$x(t) \geq c \int_{t_0}^{t} [\eta(t,0) - \eta(t,-r)]dt + c.$$

This implies the infinite integral exists.

Conversely, if φ in C, $|\varphi| < \delta$, then $x(t) = x(t_0,\varphi)(t)$ satisfies

$$|x(t)| \leq \delta e^{\int_{t_0}^{t} [\eta(\tau,0)-\eta(\tau,-r)]d\tau}$$

and the convergence of the infinite integral is sufficient for stability at t_0.

From the practical point of view, it does not seem to be of significance to consider systems for which $x = 0$ is stable at t_0 but not stable for $t_1 \geq t_0$. Therefore, in the following, we will always concern ourselves with stability according to the following definitions.

Definition 10.2. (a) The solution $x = 0$ of (3.1) is called stable if it is stable for every $t_0 \geq 0$.

(b) The solution $x = 0$ of (3.1) is called asymptotically stable if it is asymptotically stable for every $t_0 \geq 0$.

(c) The solution $x = 0$ of (3.1) is called uniformly stable if it is stable and the number δ in the Definition 10.1 (a) does not depend upon t_0.

(d) The solution $x = 0$ of (3.1) is called uniformly asymptotically stable if it is uniformly stable and for every $\eta > 0$ and every $t_0 \geq 0$, there is a $T(\eta)$, independent of t_0, and an $H_0 > 0$ independent of η and t_0, such that φ in C, $|\varphi| < H_0$ implies

$$|x_t(t_0,\varphi)| < \eta \quad \text{for} \quad t \geq t_0 + T(\eta).$$

If $y(t)$ is any solution of (3.1), then y is said to be stable if the solution $z = 0$ of the system

$$\dot{z}(t) = f(t,z_t + y_t) - f(t,y_t)$$

is stable. The other concepts are defined in a similar manner.

11. SUFFICIENT CONDITIONS FOR STABILITY OF GENERAL SYSTEMS

In this section, we give sufficient conditions for stability of the solution $x = 0$ of (3.1) and illustrate the results with examples. If $V: R^+ \times C_H \to R$ is continuous we let

$$\dot{V}(t,\varphi) = \overline{\lim_{h \to 0^+}} \frac{1}{h}[V(t+h, x_{t+h}(t,\varphi)) - V(t,\varphi)]$$

where $x_{t+h}(t,\varphi)$ is the solution of (3.1) through (t,φ). $\dot{V}(t,\varphi)$ is the upper right hand derivate of $V(t,\varphi)$ along the solutions of (3.1).

<u>Theorem 11.1.</u> Suppose f takes closed bounded sets of $R^+ \times C_H$ into closed bounded sets of R^n. Suppose $u(s)$, $v(s)$, $w(s)$ are continuous functions for s in $[0,H)$, $u(s)$, $v(s)$ positive and nondecreasing for $s \neq 0$, $u(0) = v(0) = 0$, $w(s)$ nonnegative, and nondecreasing. If there is a continuous function V: $R^+ \times C_H \to R$ such that

$$u(|\varphi(0)|) \leq V(t,\varphi) \leq v(|\varphi|)$$

$$\dot{V}(t,\varphi) \leq -w(|\varphi(0)|)$$

then the solution $x = 0$ of (3.1) is uniformly stable. If, in addition, $w(s) > 0$ for $s > 0$, $w(s)$ nondecreasing, then the solution $x = 0$ of (3.1) is uniformly asymptotically stable.

<u>Proof.</u> For any $\varepsilon > 0$, there is a $\delta = \delta(\varepsilon)$, $0 < \delta < \varepsilon$, such that $v(\delta) < u(\varepsilon)$. If $\varphi \in C_\delta$, $t_0 \geq 0$, then $\dot{V}(t, x_t(t_0,\varphi)) \leq 0$ for all $t \geq t_0$ and the inequalities on $V(t,\varphi)$ imply

$$\begin{aligned} u(|x(t_0,\varphi)(t)|) &\leq V(t, x_t(t_0,\varphi)) \\ &\leq V(t_0,\varphi) \leq v(\delta) < u(\varepsilon), \quad t \geq t_0. \end{aligned}$$

Therefore, $|x(t_0,\varphi)(t)| < \varepsilon$, $t \geq t_0$. Since $|\varphi| < \delta < \varepsilon$, this proves uniform stability.

For $\varepsilon = 1$, choose $\delta_0 = \delta(1)$ as the above constant for uniform stability. For any $\varepsilon > 0$, we wish to show there is a $T(\delta_0,\varepsilon) > 0$ such that any solution $x(t_0,\varphi)$ of (3.1) with $|\varphi| < \delta_0$ satisfies $|x_t(t_0,\varphi)| < \varepsilon$ for $t \geq t_0 + T(\delta_0,\varepsilon)$. Let $\delta = \delta(\varepsilon)$ be the above constant for uniform stability. Suppose that a solution $x = x(t_0,\varphi)$, $|\varphi| < \delta_0$ satisfies $|x_t| \geq \delta$, $t \geq t_0$. Then there exists a sequence $\{t_k\}$ such that

$$t_0 + (2k-1)r \leq t_k \leq t_0 + 2kr, \quad k = 1,2,\ldots,$$

and

$$|x(t_k)| \geq \delta.$$

By the assumption on f, there exists a constant L such that $|\dot{x}(t)| < L$ for all $t \geq t_0$. Therefore, on the intervals $t_k - \delta/2L \leq t \leq t_k + \delta/2L$, we have $|x(t)| > \delta/2$, and hence $|x_t| > \delta/2$. Therefore,

$$\dot{V}(t,x_t) \leq -w(\tfrac{\delta}{2}), \quad t_k - \frac{\delta}{2L} \leq t \leq t_k + \frac{\delta}{2L}.$$

By taking a large L, if necessary, we can assume that these intervals do not overlap, and hence

$$V(t_k,x_{t_k}) - V(t_0,\varphi) \leq -w(\tfrac{\delta}{2})\ \tfrac{\delta}{L}(k-1).$$

Let $K(\delta_0,L)$ be the smallest integer $\geq v(\delta_0)/((\delta/L)w(\delta/2))$. If $k > 1 + K(\delta_0,L)$, then

$$V(t_k,x_{t_k}) < v(\delta_0) - w(\tfrac{\delta}{2})\ \frac{\delta}{L}\ \frac{v(\delta_0)}{w(\frac{\delta}{2})\ \frac{\delta}{L}} = 0,$$

which is a contradiction. Therefore, at some t_1, such that $t_0 \le t_1 \le t_0 + 2rK(\delta_0, L)$ we have $|x_{t_1}| < \varepsilon$. This proves the theorem.

Let us consider a possible method of construction of a particular Lyapunov function for the equation

$$\dot{x}(t) = Ax(t) + Bx(t-r), \quad r > 0,$$

where A, B are constant matrices. Suppose A is an asymptotically stable matrix and choose C such that $C > 0$, $A'C + CA = -D < 0$. If E is a positive definite matrix and

$$V(\varphi) = \varphi'(0)C\varphi(0) + \int_{-r}^{0} \varphi'(\theta)E\varphi(\theta)d\theta$$

then $\nu|\varphi(0)|^2 \le V(\varphi) \le K|\varphi|$ for some positive ν, K. Furthermore,

$$V(\varphi) = -\varphi'(0)D\varphi(0) + 2\varphi'(0)CB\varphi(-r)$$
$$+ \varphi'(0)E\varphi(0) - \varphi'(-r)E\varphi(-r).$$

This is a bilinear form in $\varphi(0)$, $\varphi(-r)$ which for appropriate choices of E and B can be made negative in $\varphi(0)$, $\varphi(-r)$. In fact, for E and B sufficiently small in norm, this is always true. To be more specific, suppose $E < D$ and

$$x'(D-E)x \ge \lambda|x|^2, \quad x'Ex \ge \mu|x|^2.$$

Then

$$\dot{V}(\varphi) \le -\lambda|\varphi(0)|^2 + 2\|CB\| \cdot |\varphi(0)| \cdot |\varphi(-r)| - \mu|\varphi(-r)|^2$$

and if $2\lambda\mu - \|CB\| > 0$, then $\dot{V}(\varphi) \le -k(|\varphi(0)|^2 + |\varphi(-r)|^2)$, $r > 0$, for a suitable positive constant k Theorem 11.1 implies uniform asymptotic stability for all r.

<u>Example 11.1.</u> Consider the scalar equation

$$\dot{x}(t) = -ax(t) - b(t)x(t-r) \tag{11.1}$$

where $a > 0$, $b(t)$ is continuous and bounded for all $t \geq 0$. If x is scalar, take $|x|$ as the absolute value of x. If

$$V(\varphi) = \frac{1}{2a}\varphi^2(0) + \mu \int_{-r}^{0} \varphi^2(\theta)d\theta,$$

where $\mu > 0$ is to be determined, then

$$\dot{V}(x_t) = -x^2(t) - \frac{b(t)}{a}x(t)x(t-r)$$
$$+ \mu x^2(t) - \mu x^2(t-r)$$

and $\dot{V}(x_t)$ is a negative definite quadratic function of $x(t)$, $x(t-r)$ uniformly in t if there is a $\theta < 1$ such that

$$\theta r(1-\mu)\mu > \frac{b^2(t)}{a^2}$$

or, for $\mu = 1/2$, $b^2(t) < a^2\theta$. But this implies $\dot{V}(x_t) \leq -k|x(t)|^2$ for some $k > 0$. The conditions of Theorem 11.1 are now satisfied and we conclude that the solution $x = 0$ of (6.1) is uniformly asymptotically stable.

In case $a = a(t)$ and $r = r(t)$, $0 \leq r(t) \leq r$, then a similar argument yields stability criteria with

$$V(t,x_t) = \psi(t)x^2(t) + \mu\int_{-r(t)}^{0} x^2(t+\theta)d\theta.$$

The condition on the parameters obtained by using this V are

$$[2a(t)\psi(t) - \mu - \psi'(t)][1 - r'(t)]\mu$$
$$> b^2(t)\psi^2(t).$$

In the above discussion of Example 11.1, the stability region obtained was independent of r and the sign of $b(t)$. If b is a constant, then the exact region of stability is indicated in the Figure 1.

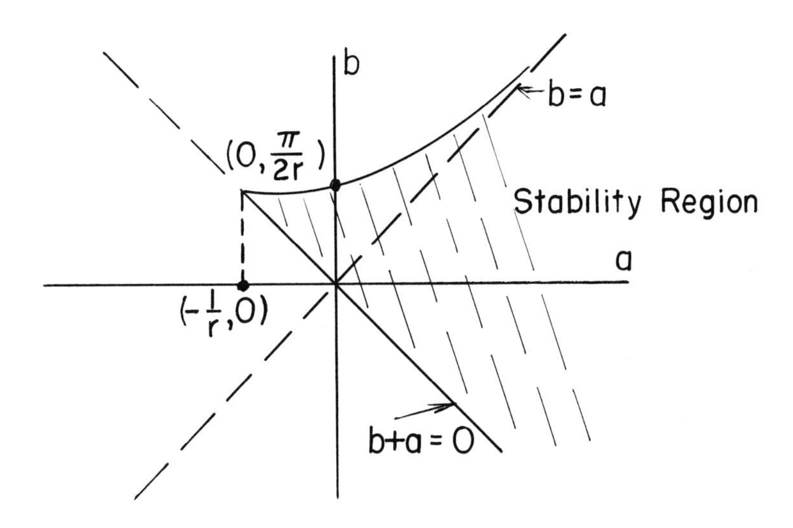

The region $|b| < a$ is the maximum region for which stability is assured for all values of r, $0 \leq r < \infty$. On the other hand, as $r \to 0$ the true region of stability for (11.1) approaches the half-plane $b + a > 0$.

We ask the following question: is it possible to use a Lyapunov function to obtain a region of stability which depends upon r and has the same qualitative structure as the true region of stability? In particular, can we use a Lyapunov function to obtain a region of stability which approaches the half-plane $b + a > 0$ as $r \to 0$?

Let α be a constant and $\beta(\theta)$ be a continuously differentiable function of θ for $-r \leq \theta \leq 0$. If

$$(11.2) \qquad V(x_t) = \frac{x^2(t)}{2|a|} + \alpha x(t) \int_{t-r}^{t} x(u)du + \int_{t-r}^{t} \beta(u-t)x^2(u)du,$$

then there are $k > a$, $K > 0$, such that

$$k|x(t)|^2 \leq V(x_t) \leq K|x_t|,$$

if

$$(11.3) \qquad \beta(\theta) > \frac{\alpha^2|a|r}{2}, \quad -r \leq \theta \leq 0.$$

Furthermore,

$$\overset{\circ}{V}(x_t) = -\int_{t-r}^{t} [\frac{\operatorname{sgn} a - \alpha - \beta(0)}{r} x^2(t) + \frac{1}{r}(\alpha + \frac{b}{|a|})x(t)x(t-r) + \frac{\beta(-r)}{r} x^2(t-r) + \alpha a x(t)x(u) + \alpha b x(t-r)x(u) + \overset{\circ}{\beta}(u-t)x^2(u)]du.$$

Consequently, if β satisfies (11.3) and the integrand is positive definite in $x(t)$, $x(t-r)$ and $x(u)$, then Theorem 11.1 will imply the uniform asymptotic stability of the solution $x = 0$ of (11.1). The necessary and sufficient conditions for the integrand to be positive definite in these variables is that $\alpha,\beta,a,b(t),c$ satisfy the following set of inequalities:

$$(11.4) \qquad \begin{aligned} \Delta_1 &\overset{\text{def}}{=} \operatorname{sgn} a - \alpha - \beta(0) > 0 \\ \Delta_2 &\overset{\text{def}}{=} [\operatorname{sgn} a - \alpha - \beta(0)]\beta(-r) - \frac{1}{4}(\alpha + \frac{b}{|a|})^2 > 0 \\ \Delta_3 &\overset{\text{def}}{=} \frac{\Delta_2}{r^2} \dot{\beta}(\theta) - \frac{\alpha^2 a^2}{4r} \end{aligned}$$

$$\cdot\left[\frac{b^2}{a^2}(\operatorname{sgn} a - \alpha - \beta(0)) - \frac{b(t)}{a}\left(\alpha + \frac{b}{|a|}\right) + \beta(-r)\right] > 0.$$

Inequalities (11.4) will be satisfied if

$$\Delta_1 > 0, \quad \Delta_2 > 0$$

(11.5)

$$\dot{\beta}(0) > \frac{r\alpha^2 b^2}{4\beta(-r)} + \frac{r\alpha^2 a^2}{4\beta(-r)\Delta_2}\left[\frac{b}{2a}\left(\alpha + \frac{b}{|a|}\right) - \beta(-r)\right]^2$$

Now suppose b is independent of t and choose

$$\alpha = \frac{2a\beta(-r)}{b} - \frac{b}{|a|}.$$

All of the above inequalities will be satisfied if

$$a + b > \frac{a|a|}{b}\left(2 + \frac{a}{b}\right)\beta(-r) + |a|\beta(0)$$

$$\dot{\beta}(\theta) > \frac{r\alpha^2}{\beta(-r)}\frac{b^2}{4}, \quad \frac{r\alpha^2}{\beta(-r)} < \frac{2}{|a|}.$$

Suppose a, b are fixed and $a + b > 0$. If we choose $\beta(-r)$ as a function of r such that $\beta(-r) \to 0$, $r/\beta(-r) \to 0$ as $r \to 0$, then $\alpha \to -b/|a|$ as $r \to 0$ and one can obviously choose $\dot{\beta}(\theta)$ as a function of r so that $\beta(0) \to 0$ as $r \to 0$ and all of the above inequalities are satisfied for r sufficiently small.

Even if b depends on t, one can obtain some results using (11.5) with a little extra effort.

As an instructive exercise, try to determine regions of stability for the equation

$$\dot{x}(t) = -ax(t) - b(t)x(t-r_1) - c(t)x(t-r_2).$$

Theorem 11.2. Suppose $f(s) > s$ for $s > 0$ is a continuous nondecreasing function and $u(s)$, $v(s)$, $w(s)$ are continuous functions for s in $[0,H]$, with $u(s)$, $v(s)$ positive and nondecreasing for $s \neq 0$, $v(0) = 0$ and $w(s)$ positive for $s \neq 0$. If there exists a function $V(t,x)$ which is defined and continuous for $t \geq -r$, $x \in R^n$, $|x| < H$, such that

$$(11.6) \qquad u(|x|) \leq V(t,x) \leq v(|x|), \quad 0 \leq |x| < H, \quad t \geq -r,$$

and

$$(11.7) \qquad \dot{V}(t,x(t)) \leq -w(|x(t)|)$$

for all $t \geq 0$ and those functions x_t in C_H such that

$$(11.8) \qquad V(\xi,x(\xi)) < f(V(t,x(t))), \quad t-r \leq \xi \leq t,$$

then the solution $x = 0$ of (3.1) is uniformly asymptotically stable.

Proof. To prove that the solution $x = 0$ is uniformly stable, let

$$\bar{V}(t,x_t) = \sup_{t-r\leq\xi\leq t} V(\xi,x(\xi)), \quad x_t \text{ in } C_H, \ t \geq 0.$$

Then $u(|x(t)|) \leq \bar{V}(t,x_t) \leq v(|x_t|)$ and conditions (11.7), (11.8) imply that $\dot{\bar{V}}(t,x_t) \leq 0$ for all $t \geq 0$, $x_t \in C_H$. Theorem 11.1 implies that the solution $x = 0$ of (3.1) is uniformly stable.

Choose $\delta_1 > 0$ so small that there is an H_1 such that $v(\delta_1) = u(H_1)$. Then $|\varphi| \leq \delta_1$ implies $|x_t(t_0,\varphi)| \leq H_1$, $V(t,x(t_0,\varphi)(t)) \leq v(\delta_1)$ for $t \geq t_0-r$. Suppose $0 < \eta \leq H_1$ is arbitrary. We need to show there is a number $T = T(\eta,\delta_1)$ such that for any $t_0 \geq 0$ and $|\varphi| \leq \delta_1$ the solution $x(t_0,\varphi)$ of (3.1) satisfies $|x_t(t_0,\varphi)| \leq \eta$, $t \geq t_0 + T + r$ or, equivalently, $V(t,x(t_0,\varphi)(t)) \leq u(\eta)$, for

$t \geq t_0 + T$. In the remainder of this proof, we let $x(t) = x(t_0,\varphi)(t)$.

From the properties of the function $f(s)$, there is a number $a > 0$ such that $f(s) - s > a$ for $u(\eta) \leq s \leq v(\delta_1)$. Let N be any positive integer such that $u(\eta) + Na \geq v(\delta_1)$, and let $\gamma = \inf_{\eta \leq s \leq H_1} w(s)$, $T = Nv(\delta_1)/\gamma$.

We now show that $V(t,x(t)) \leq u(\eta)$ for all $t \geq t_0 + T + (N-1)r$. First we show that $V(t,x(t)) \leq u(\eta) + (N-1)a$ for $t \geq t_0 + v(\delta_1)/\gamma$. If $V(t,x(t)) > u(\eta) + (N-1)a$ for $t_0 - r \leq t < t_0 + v(\delta_1)/\gamma$, then, since $V(t,x(t)) \leq u(\eta) + Na$ for all $t \geq t_0$, it follows that

$$f(V(t,x(t))) > V(t,x(t)) + a \geq u(\eta) + Na \geq V(s,x(s)),$$

$$t_0 - r \leq s \leq t_0 + \frac{v(\delta_1)}{\gamma}$$

and

$$\dot{V}(t,x(t)) \leq -w(|x(t)|) \leq -\gamma.$$

Consequently,

$$V(t,x(t)) \leq V(t_0,x(t_0)) - \gamma(t-t_0)$$
$$\leq v(\delta_1) - \gamma(t-t_0) \leq 0$$

if $t \geq t_0 + v(\delta_1)/\gamma$, which implies that $V(t,x(t)) \leq u(\eta) + (N-1)a$ at $t_1 = t_0 + v(\delta_1)/\gamma$. But, this implies $V(t,x(t)) \leq u(\eta) + (N-1)a$ for all $t \geq t_0 + v(\delta_1)/\gamma$, since $\dot{V}(t,x(t))$ is negative [by (11.6) and (11.7)] when $V(t,x(t)) = u(\eta) + (N-1)a$.

Now, let $\bar{t}_j = jv(\delta_1)/\gamma + (j-1)r$, $j = 0,1,2,\ldots,N$, and assume that for some integer $k \geq 1$, in the interval

$$\bar{t}_{k-1} \le t - t_0 \le t_k, \quad t_0 = 0, \ t_j = \frac{jv(\delta_1)}{\gamma},$$

$$j = 1,2,\ldots,$$

we have

$$u(\eta) + (N-k)a \le V(t,x(t)) \le u(\eta) + (N-k+1)a.$$

By the same type of reasoning as above, we have

$$\dot{V}(t,x(t)) \le -\gamma, \quad \bar{t}_{k-1} \le t - t_0 \le \bar{t}_k$$

and

$$V(t,x(t)) \le V(t_0 + \bar{t}_{k-1}, x(t_0 + \bar{t}_{k-1})) - \gamma(t-t_0-\bar{t}_{k-1})$$

$$\le v(\varepsilon_1) - \gamma(t-t_0-\bar{t}_{k-1}) \le 0$$

if $t-t_0 - \bar{t}_{k-1} \ge v(\delta_1)/\gamma$. Consequently, $V(t_0 + \bar{t}_k, x(t_0 + \bar{t}_k)) \le u(\eta) + (N-k)a$, and, finally, $V(t,x(t)) \le u(\eta) + (N-k)a$ for all $t \ge t_0 + \bar{t}_k$. This completes the induction and we have $V(t,x(t)) \le u(\eta)$ for all $t \ge t_0 + Nv(\delta_1)/\gamma + (N-1)r$. This proves Theorem 11.2.

<u>Example 11.2.</u> If we again consider equation (11.1) and let $V(x) = x^2/2a$, then

$$\dot{V}(x(t)) = -x^2(t) + \frac{b(t)}{a}x(t)x(t-r).$$

If $f(s) = qs$, $q > 1$, then

$$\dot{V}(x(t)) \le -(1 - \frac{q|b(t)|}{a})x^2(t) \le -\delta x^2(t)$$

if $q|b(t)| < a(1-\delta)$ and $x^2(\xi) < qx^2(t)$, $t-r \leq \xi \leq t$. Consequently, Theorem 11.2 implies the solution $x = 0$ of (11.1) is uniformly asymptotically stable. Notice that $r = r(t)$ does not change the above argument if $r(t)$ is continuous and bounded.

If we use the same $V(x)$ as above, then a similar argument shows that the zero solution of

$$\dot{x}(t) = -ax(t) - \sum_{j=1}^{n} b_j(t)x(t-r_j(t)),$$

is uniformly asymptotically stable for all bounded continuous $r_j(t)$, if $\sum_{j=1}^{n}|b_j(t)| < qa$, $a > 0$, $0 < q < 1$.

Example 11.3. Consider the first order equation

$$\dot{x}(t) = f(x(t-\gamma(t)),t), \quad 0 \leq \gamma(t) \leq r, \quad f(0,t) = 0 \tag{11.8}$$

where $\gamma(t)$ is a continuous function of t and $f(x,t)$ is a continuous function of x,t for $t \geq 0$, $-\infty < x < \infty$, has a continuous partial derivative such that $|\partial f(x,t)/\partial x| < L$, $t \geq 0$, $-\infty < x < \infty$. For $t \geq 2r$ we can rewrite equation (11.8) as

$$\begin{aligned} \dot{x}(t) &= f(x(t),t) - [f(x(t),t) - f(x(t-\gamma(t)),t)] \\ &= f(x(t),t) - \int_{t-\gamma(t)}^{t} \frac{\partial f}{\partial x}(x(\theta),t)f(x(\theta-\gamma(\theta)),\theta)d\theta. \end{aligned} \tag{11.9}$$

For $V(\varphi) = \varphi^2(0)$, we have

$$\begin{aligned} \dot{V}(x_t) &= 2x(t)f(x(t),t) \\ &\quad -2\int_{t-\gamma(t)}^{t} x(t)\frac{\partial f}{\partial x}(x(\theta),t)f(x(\theta-\gamma(\theta)),\theta)d\theta \end{aligned}$$

$$\leq 2x(t)f(x(t),t) + 2L^2 \int_{t-\gamma(t)}^{t} |x(t)x(\theta-\gamma(\theta))|\,d\theta$$

$$\leq 2x(t)f(x(t),t) + 2L^2\gamma(t)|x(t)|\cdot|x_{t-\gamma(t)}| \quad t \geq 2r.$$

Consequently, if we consider the set of all $x(t)$ such that

$$qx^2(t-\xi) \leq x^2(t), \quad 0 \leq \xi \leq 2r, \quad 0 < q < 1,$$

then

$$V(x_t) \leq 2[\frac{f(x(t),t)}{x(t)} + L^2\gamma(t)]x^2(t) < -2\mu x^2(t)$$

if $f(x,t)/x + L^2\gamma(t) < -\mu < 0$ for $\mu > 0$, $t \geq 0$, $x \in (-\infty,\infty)$, and Theorem 11.2 implies the origin is globally asymptotically stable.

Example 11.4. Consider the equation

$$\dot{x}(t) = Ax(t) + Bx(t-\tau) \tag{11.10}$$

where A,B are matrices (A constant, B may be a bounded variable matrix) and $\tau = \tau(t)$, $0 \leq \tau(t) \leq r$, is continuous. If $V(x) = x'Dx$, where C is positive definite, then

$$\dot{V} = x'(t)(D'A + AD)x(t) + 2x'(t)Bx(t-\tau).$$

If $\qquad V(x(\xi)) < qV(x(t)), \quad q > 1, \quad t-r \leq \xi \leq t$ implies

$$\dot{V} \leq -\omega(|x(t)|)$$

$\omega(s) > 0$ if $s > 0$, then Theorem 11.2 implies the solution $x = 0$ of (11.10) is uniformly asymptotically stable.

The difficulty in obtaining results along this line arises from attempting to estimate $\dot{V}$ for the restricted class of initial curves satisfying $V(x(\xi)) < qV(x(t))$, $t-r \leq \xi \leq t$. Furthermore, there are numerous directions in which one may proceed. In particular, one may wish to obtain stability conditions which are independent of τ or conditions which depend upon τ. In the first case, one must obviously have the zero solution of

$$\dot{x}(t) = (A+B)x(t) \tag{11.11}$$

asymptotically stable. By an appropriate change of coordinates, one can take $V(x) = x'x$ and be assured that $\dot{V}$ is a negative definite function. In these new coordinates, $\dot{V}$ along solutions of (11.10) is

$$\begin{aligned}\dot{V} = x'(t)[(A+B) + (A+B)']x(t) \\ + 2x'(t)Bx(t-\tau) - x'(t)(B+B')x(t)\end{aligned}$$

and one can estimate $\dot{V}$ along curves satisfying $V(x(\xi)) < qV(x(t))$ (or, equivalently, $|x(\xi)| < q|x(t)|$), $q > 1$, $t - \tau \leq \xi \leq t$ in the following way. Since $(A+B) + (A'+B')$ is negative definite there is a $\lambda > 0$ such that

$$\begin{aligned}\dot{V} &\leq -\lambda|x(t)|^2 + 2\tau|B|\cdot|x(t)|^2 + |B+B'|\cdot|x(t)|^2 \\ &= -[\lambda - 2\tau|B| - |B+B'|]|x(t)|^2.\end{aligned}$$

Consequently, if $2\tau|B| + |B+B'| < \lambda$, then the solution $x = 0$ of (11.10) is uniformly asymptotically stable. Razumikhin [1] has carried out this type of procedure for a second order system

$$\ddot{x}(t) + a\dot{x}(t) + bx(t) + cx(t-\tau) = 0$$

to obtain estimates on a,b,c ensuring asymptotic stability independent of τ.

To obtain estimates dependent on τ, we observe that solutions of (11.10) are differentiable in t for $t \geq \tau$ and, therefore, from the point of view of stability, we may as well assume continuously differentiable initial values. Consequently,

$$x(t-\tau) = x(t) - \int_{-\tau}^{0} \dot{x}(t-\xi)d\xi$$

$$= x(t) - \int_{-\tau}^{0} [Ax(t-\xi) + Bx(t-\tau-\xi)]d\xi.$$

Using the same argument as above, one can obtain estimates involving the magnitude of τ which ensure asymptotic stability.

12. SUFFICIENT CONDITIONS FOR INSTABILITY

In this section, we give a sufficient condition for the instability of the solution $x = 0$ of (2.1) and give some examples to illustrate the result.

<u>Theorem 12.1.</u> Suppose $V(\varphi)$ is a continuous bounded scalar function on C_H. If there exist a γ, $0 < \gamma < H$ and an open set U in C such that

(i) $V(\varphi) > 0$ on U, $V(\varphi) = 0$ on the boundary of U,

(ii) 0 belongs to the closure of $U \cap C_\gamma$,

(iii) $V(\varphi) \leq u(|\varphi(0)|)$ on $U \cap C_\gamma$,

(iv) $\dot{V}^*(\varphi) \geq w(|\varphi(0)|)$ on $[0,\infty) \times U \cap C_\gamma$,

$$\dot{V}^*(\varphi) = \varliminf_{h \to 0^+} \frac{1}{h}[V(x_{t+h}(t,\varphi)) - V(\varphi)],$$

where $u(s)$, $v(s)$ are continuous, increasing and positive for $s > 0$, then the solution $x = 0$ of (2.1) is unstable. More specifically, each solution $x_t(t_0,\varphi)$ of (2.1) with initial function φ in $U \cap C_\gamma$ at t_0 must reach the boundary of C_γ in a finite time.

<u>Proof.</u> Suppose $\varphi_0 \in U \cap C_\gamma$, $t_0 \geq 0$. Then $V(\varphi_0) > 0$. By hypothesis (iii), $|\varphi_0(0)| \geq u^{-1}(V(\varphi_0))$ and (iii) and (iv) imply $x_t = x_t(t_0,\varphi_0)$ satisfies $|x(t)| \geq u^{-1}(V(x_t)) \geq u^{-1}(V(\varphi_0))$ as long as $x_t \in U \cap C_\gamma$. Consequently

$$\dot{V}^*(x_t) \geq w(|x(t)|) \geq w(u^{-1}(V(\varphi_0))) > 0 \quad \text{if} \quad x_t \in U \cap C_\gamma.$$

If we let $\eta = w(u^{-1}(V(\varphi_0)))$, then this implies

$$V(x_t) \geq V(\varphi_0) + \eta(t-t_0)$$

as long as $x_t \in U \cap C_\gamma$. Hypotheses (i) and (iv) imply that x_t cannot leave $U \cap C_\gamma$ by crossing the boundary of U. Since $V(\varphi)$ is bounded on $U \cap C_\gamma$ this implies there must be a t_1 such that $x_{t_1} \in \overline{C}_\gamma$. This proves the last assertion of the theorem. But hypothesis (ii) implies that in each neighborhood of the origin of C, there are φ_0 in $U \cap C_\gamma$ and the instability follows, completing the proof of the theorem.

Example 12.1. Consider the equation

$$\dot{x}(t) = -ax(t) - bx(t-r),$$

where $a + b < 0$ and r is any positive constant. We wish to prove by use of Lyapunov functions that the solution $x = 0$ of this equation is unstable. The exact region of stability for this equation is shown in Figure 1 of Section 11. The region $a + b < 0$ is the interior of the intersections of the instability regions as a function of r.

If F is any given function and

$$V(x_t) = \frac{x^2(t)}{2} - \frac{1}{2}\int_{t-r}^{t} F(t-u)[x(u) - x(t)]^2 du,$$

then it is easily seen that

$$\begin{aligned}\dot{V}^*(x_t) &= \dot{V}(x_t) \\ &= -(a+b)x^2(t) - b[x(t-r) - x(t)]x(t) \\ &\qquad + \frac{1}{2} F(r)[x(t-r) - x(t)]^2 \\ &\quad - \frac{1}{2}\int_{t-r}^{t} \dot{F}(t-u)[x(u) - x(t)]^2 du \\ &\quad + \int_{t-r}^{t} F(t-u)[x(u) - x(t)]\cdot[-(a+b)x(t) - b\{x(t-r)-x(t)\}]du.\end{aligned}$$

If the expression for $\dot{V}$ is written as an integral from $[t-r,t]$, then the integrand will be a positive definite quadratic form in $x(t)$, $[x(t-r) - x(t)]$, $[x(u) - x(t)]$ if the following inequalities are satisfied

$$a + b < 0$$

$$\Delta_2 \overset{\text{def}}{=} -\frac{(a+b)}{2} F(r) - \frac{b^2}{4} > 0$$

$$\Delta_3 \overset{\text{def}}{=} \frac{\Delta_2}{r^2}\left(-\frac{1}{2}\dot{F}(\theta)\right) - \frac{(a+b)^2}{8r} F^3(\theta) > 0, \quad 0 \leq \theta \leq r.$$

If $a + b < 0$, it is clear these inequalities can be satisfied by a continuously differentiable positive function $F(\theta)$, $0 \leq \theta \leq r$. Consequently, there exists a positive number q such that

$$\dot{V}^*(\varphi) \geq qr\varphi^2(0), \quad V(\varphi) \leq \varphi^2(0)/2,$$

for all φ in C. If

$$U = \{\varphi \text{ in } C: \varphi^2(0) > \int_{-r}^{0} F(-\theta)[\varphi(\theta) - \varphi(0)]^2 d\theta\},$$

then U satisfies (i) and (ii) of Theorem 12.1 and the solution $x = 0$ is unstable.

Notice the same conclusions for this example are valid if a,b are functions of t provided that a,b are bounded and $a(t) + b(t) < \delta < 0$ for all t.

Example 12.2. Consider the equation

$$\dot{x}(t) = ax^3(t) + b(t)x^3(t-r)$$

with $a > 0$, $|b(t)| < qa$, $0 < q < 1$ arbitrary. For

$$V(\varphi) = \frac{\varphi^4(0)}{4a} - \frac{1}{2}\int_{-r}^{0} \varphi^6(\theta)d\theta \leq \frac{\varphi^4(0)}{4a}$$

we have

$$\dot{V}^*(\varphi) = \dot{V}(\varphi) = \frac{1}{2}\varphi^6(0) + \frac{b(t)}{a}\varphi^3(0)\varphi^3(-r) + \frac{1}{2}\varphi^6(-r)$$

$$\geq \frac{1}{2}(1-q)(\varphi^6(0) + \varphi^6(-r))$$

$$\geq \frac{1}{2}(1-q)\varphi^6(0)$$

and if $U = \{\varphi \in C: \varphi^4(0) > \int_{-r}^{0} \varphi^6(\theta)d\theta\}$, then the same argument as before shows that $x = 0$ is an unstable solution of this equation.

If $a < 0$, $|b(t)| < q|a|$, one can choose

$$V(\varphi) = \frac{\varphi^4(0)}{4|a|} + \frac{1}{2}\int_{-r}^{0} \varphi^6(\theta)d\theta$$

and use Theorem 11.1 to prove the zero solution is stable.

Notice that the same V functionals may be used to show that the zero solution of $\dot{x}(t) = ax^3(t) + b(t)x^4(t-r)$ is stable or unstable according as $a < 0$ or > 0, regardless of the bound on $b(t)$. One simply must operate in a sufficiently small neighborhood of the origin.

13. STABILITY IN AUTONOMOUS SYSTEMS

Consider the equation

$$\dot{x}(t) = f(x_t) \tag{13.1}$$

where $f: C \to R$ is continuous, takes closed bounded sets into bounded sets and solutions of (13.1) depend continuously upon the initial data. We denote by $x(\varphi)$ the solution through $(0,\varphi)$.

If $V: C \to R$ is a continuous function, we define

$$\dot{V}(\varphi) = \dot{V}_{(13.1)}(\varphi) = \overline{\lim_{h \to 0^+}} \frac{1}{h} [V(x_h(\varphi)) - V(\varphi)].$$

<u>Definition 13.1.</u> We say $V: C \to R$ is a <u>Lyapunov function on a set</u> G <u>in</u> C relative to system (13.1) if V is continuous on $\bar{G}$, the closure of G, and $\dot{V} \leq 0$ on G. Let

$$S = \{\varphi \in \bar{G}: \dot{V}(\varphi) = 0\}$$

M = largest set in S which is invariant with respect to (13.1).

<u>Theorem 13.1.</u> If V is a Lyapunov function on G and $x_t(\varphi)$ is a bounded solution of (13.1) which remains in G, then $x_t(\varphi) \to M$ as $t \to \infty$.

<u>Theorem 13.2.</u> If V is a Lyapunov function on $U_\ell = \{\varphi \in C: V(\varphi) < \ell\}$ and there is a constant $K = K(\ell)$ such that φ in U_ℓ implies $|\varphi(0)| < K$, then any solution $x_t(\varphi)$ of (13.1) with φ in $U_\ell \to M$ as $t \to \infty$.

<u>Proof of Theorem 13.1.</u> If $|x_t(\varphi)| < K$, $x_t(\varphi) \in G$, $t \geq 0$, then $\{x_t(\varphi)\}$ belongs to a compact set of C and has a nonempty ω-limit set, $\omega(\gamma^+(\varphi))$. Thus, $V(x_t(\varphi))$ is nonincreasing, bounded below and must approach a limit c as $t \to \infty$. Since V is continuous on $\bar{G}$, $V(\psi) = c$ for ψ in $\omega(\gamma^+(\varphi))$. Since $\omega(\gamma^+(\varphi))$ is invariant $\dot{V}(\psi) = 0$ on $\omega(\gamma^+(\varphi))$. Since every solution approaches its ω-limit set,

this proves Theorem 13.1.

Proof of Theorem 13.2. If φ is in U_ℓ, $\dot{V} \leq 0$ on U_ℓ, then $x_t(\varphi) \in U_\ell$, $t \geq 0$. Also $|x(\varphi)(t)| \leq K$ if $t \geq 0$ which implies $x_t(\varphi)$ bounded. Now use Theorem 13.1.

Corollary 13.1. Suppose $V: C \to R$ is continuous and there exist nonnegative functions $a(r)$, $b(r)$ such that $a(r) \to \infty$ as $r \to \infty$

$$a(|\varphi(0)|) \leq V(\varphi)$$

$$\dot{V}(\varphi) \leq -b(|\varphi(0)|).$$

Then every solution of (13.1) is bounded. If, in addition $b(r)$ is positive definite, then every solution approaches zero as $t \to \infty$.

Proof. The boundedness of solutions is immediate since $a(r) \to \infty$ and $|x(t)|$ bounded for $t \geq 0$ implies x_t bounded for $t \geq 0$. If b is positive definite, the conditions of Theorem 13.2 are satisfied for any ℓ. Furthermore, $S = \{\varphi: \varphi(0) = 0\}$, $M = \{0\}$.

Theorem 13.3. (Instability). Suppose zero belongs to the closure of an open set U in C and N is an open neighborhood of zero in C. Assume that

(i) V is a Lyapunov function on $G = N \cap U$,

(ii) $M \cap \bar{G}$ is either the empty set or zero,

(iii) $V(\varphi) < \eta$ on G when $\varphi \neq 0$,

(iv) $V(0) = \eta$ and $V(\varphi) = \eta$ when $\varphi \in \partial G \cap N$.

If N_0 is a bounded neighborhood of zero properly contained in N, then $\varphi \neq 0$ in $G \cap N_0$ implies there exists a τ such that $x_\tau(\varphi) \in \partial N_0$.

Proof. Suppose $\varphi \in G \cap N_0$. Then $V(x_t(\varphi)) \leq V(\varphi) < \eta$ for all $t \geq 0$ as long as

$x_t(\varphi)$ remains in $N_0 \cap G$. If $x_t(\varphi)$ remains in the bounded set $N_0 \cap G$ for all $t \geq 0$, then the ω-limit set $\omega(\gamma^+(\varphi)) \subset N_0 \cap G$. Also, $\omega(\gamma^+(\varphi))$ is invariant. Hypothesis (ii) implies $\omega(\gamma^+(\varphi)) = \{0\}$. On the other hand, $V(0) = \eta$. This is a contradiction. Therefore, there is a $\tau > 0$ such that $x_\tau(\varphi) \in \partial(N_0 \cap G)$. Condition (iv) implies $x_\tau(\varphi) \in \partial N_0$ and the theorem is proved.

14. AN EXAMPLE OF LEVIN AND NOHEL

Suppose $n = 1$ and

$$f(\varphi) = -\int_{-r}^{0} a(-\theta)g(\varphi(\theta))d\theta$$

where

$$G(x) \overset{\text{def}}{=} \int_{0}^{x} g(s)ds \to \infty \quad \text{as} \quad |x| \to \infty \tag{14.1}$$

$$a(r) = 0,\ a(t) \geq 0,\ \dot{a}(t) \leq 0,\ \ddot{a}(t) \geq 0, \quad 0 \leq t \leq r, \tag{14.2}$$

are continuous. We consider the special case of (13.1) given by

$$\begin{aligned} \dot{x}(t) &= -\int_{-r}^{0} a(-\theta)g(x(t+\theta))d\theta \\ &= -\int_{t-r}^{t} a(t-u)g(x(u))du. \end{aligned} \tag{14.3}$$

Any solution of (14.3) satisfies

$$\ddot{x}(t) + a(0)g(x(t)) = -\int_{t-r}^{t} \dot{a}(t-u)g(x(u))du \tag{14.4}$$

or

$$\begin{aligned} \ddot{x}(t) + a(0)g(x(t)) &= -\dot{a}(r) \int_{t-r}^{t} g(x(t+\theta))d\theta \\ &+ \int_{-r}^{0} \ddot{a}(-\theta)\left(\int_{\theta}^{0} g(x(t+u))du\right)d\theta. \end{aligned} \tag{14.5}$$

Equation (14.4) is the model of a special type of circulating fuel nuclear reactor where x represents the neutron density. It can also serve as a one-dimensional model in viscoelasticity where x is the strain and a is the relaxation function.

If we define $V: C \to R$ by the relation

$$V(\varphi) = G(\varphi(0)) - \frac{1}{2}\int_{-r}^{0} \dot{a}(-\theta)[\int_{\theta}^{0} g(\varphi(s))ds]^2 d\theta$$

then the derivative of V along solutions of (14.3) is given by

$$\dot{V}(\varphi) = \frac{1}{2}\dot{a}(r)[\int_{-r}^{0} g(\varphi(\theta))d\theta]^2 - \frac{1}{2}\int_{-r}^{0} \ddot{a}(-\theta)[\int_{\theta}^{0} g(\varphi(s))ds]^2 d\theta.$$

Since the hypotheses on a imply that $\dot{V}(\varphi) \leq 0$, it follows from Corollary 13.1 that all solutions are bounded.

Let us now apply Theorem 13.1 to this equation. If for any $s \in [0,r]$, we let

$$H_s(\varphi) = \int_{-s}^{0} g(\varphi(\theta))d\theta \tag{14.6}$$

then the set S of Theorem 13.1 is

$$S = \{\varphi \in C: H_r(\varphi) = 0 \text{ if } \dot{a}(r) \neq 0, H_s(\varphi) = 0 \text{ if } \ddot{a}(s) \neq 0\}.$$

From (14.5), the largest invariant set M of (14.3) in S satisfies

$$\begin{aligned} M \subset \{&\text{set in } C \text{ generated by bounded solutions } x \text{ of} \\ &\text{the ordinary differential equation} \\ &\ddot{x} + a(0)g(x) = 0 \\ &\text{for which} \\ &H_r(x_t) = 0,\ t \in (-\infty,\infty) \text{ if } \dot{a}(r) \neq 0 \\ &H_s(x_t) = 0,\ t \in (-\infty,\infty) \text{ if } \ddot{a}(s) \neq 0\}. \end{aligned} \tag{14.7}$$

If $\dot{a}(r) \neq 0$, x satisfying (14.7) is bounded and $H_r(x_t) = 0$ for $t \in (-\infty,\infty)$, then $\dot{x}(t) = \dot{x}(t-r)$ for all t. Therefore $x(t) = kt + $ (a periodic function of period r) and boundedness of x implies $x(t) = x(t+r)$ for all t. If there is an s_0 such that $\ddot{a}(s_0) \neq 0$, then there is an interval I_{s_0} containing s_0 such that $\ddot{a}(s) \neq 0$ for $s \in I_{s_0}$. If x satisfies (14.7), is bounded and $H_s(x_t) = 0$ for $-\infty < t < \infty$, $s \in I_{s_0}$, then $\dot{x}(t)$ is periodic of period s for every $s \in I_{s_0}$. Therefore, $\dot{x}(t)$ is constant and boundedness of x implies x is constant.

Theorem 14.1. If system (14.3) satisfies (14.1) and (14.2) and g has isolated zeros, then

(i) if there is an s such that $\ddot{a}(s) > 0$, then for any $\varphi \in C$, the ω-limit set $\omega(\varphi)$ of the orbit through φ is an equilibrium point of (14.3); that is, a zero of g;

(ii) if $\ddot{a}(s) \equiv 0$, $a \not\equiv 0$ (that is, a is linear), then for any $\varphi \in C$, the ω-limit set $\omega(\varphi)$ of the orbit through φ is a single periodic orbit of period r generated by a solution of (14.7).

Proof. (i) The remarks preceding the theorem imply $\omega(\varphi)$ is an equilibrium point and, therefore, must be a zero of g. Since $\omega(\varphi)$ is connected and the zeros of g are isolated, we have the result in (i).

(ii) Suppose $\ddot{a}(s) \equiv 0$ and choose $a(s) = (r-s)/r$. If x is a solution of (14.7) of period r, then

$$-\int_{t-r}^{t} \frac{r-(t-u)}{r} g(x(u))du = \int_{t-r}^{t} \frac{r-(t-u)}{r} \ddot{x}(u)du$$

$$= \dot{x}(u) \frac{r-(t-u)}{r}\Bigg|_{t-r}^{t} - \int_{t-r}^{t} \frac{1}{r} \dot{x}(u)du$$

$$= \dot{x}(t)$$

that is, x is a solution of (14.3). From the remarks preceding the theorem, this implies M consists of the periodic solutions of (14.7) of period r.

We first prove that if $\omega(\varphi)$ contains an equilibrium point c of (14.3), then $\omega(\varphi) = c$. We know that $\omega(\varphi)$ is a closed connected set and must be the union of r-periodic orbits of (14.7). If c is not a local minimum of $G(x)$, then the nature of the orbits of (14.7) in the $(x,\dot{x})$-plane implies there can be no r-periodic orbits in $\omega(\varphi)$ except c. If c is a local minimum of $G(x)$, then

$$V(\varphi) - G(c) = G(\varphi(0)) - G(c) + \frac{1}{2r}\int_{-r}^{0}\left[\int_{\theta}^{0} g(\varphi(s))ds\right]^2 d\theta > 0$$

for $\varphi \not\equiv c$ in a neighborhood of c. Since $V(x_t(\varphi))$ = constant for $\psi \in \omega(\varphi)$, it follows that $\omega(\varphi) = c$.

Therefore, assume $\omega(\varphi)$ contains no constant solutions of (14.7). Since the solutions of (14.7) must lie on the curves

$$\frac{\dot{x}^2}{2} + G(x) = \text{constant},$$

it follows that any periodic orbit must be symmetrical with respect to the x-axis. Let $u(0,\alpha) = \alpha$, $\dot{u}(0,\alpha) = 0$, $u(t,\alpha) \geq \alpha$, nonconstant periodic solution of (14.7) of least period p. Then there is an integer m such that $mp = r$. If there is an interval of periodic orbits in $\omega(\varphi)$, then p is independent of α in this interval. In fact, $p = p(\alpha)$ is continuous and, therefore, $m = m(\alpha)$ is continuous. But, m is an integer and, therefore, must be independent of α. Also,

$$
\begin{aligned}
V(u_t(\alpha)) &= V(u_0(\alpha)) \\
&= G(\alpha) + \frac{1}{2r}\int_{-r}^{0}\left[\int_{\theta}^{0} g(u(s))ds\right]^2 d\theta \\
&= G(\alpha) + \frac{1}{2r}\int_{-r}^{0} \dot{u}^2(\theta,\alpha)d\theta \\
&= G(\alpha) + \frac{1}{2mp}\int_{-mp}^{0} \dot{u}^2(\theta,\alpha)d\theta \\
&= G(\alpha) + \frac{1}{2p}\int_{-p}^{0} \dot{u}^2(\theta,\alpha)d\theta \\
&= G(\alpha) + \frac{1}{p}\int_{0}^{p/2} \dot{u}^2(\theta,\alpha)d\theta \\
&= G(\alpha) + \frac{2}{p}\int_{0}^{p/2} [G(\alpha) - G(u(\theta,\alpha))]d\theta \\
&= G(\alpha) + \frac{\sqrt{2}}{p}\int_{\alpha}^{\gamma(\alpha)} [G(\alpha) - G(\tau)]^{1/2}d\tau
\end{aligned}
$$

where $\gamma(\alpha) = u(\alpha, p/2)$. On the other hand, we must have $G'(\alpha) < 0$ in order to have a periodic solution in the first place and, also, $\int_{\alpha}^{\gamma(\alpha)} [G(\alpha) - G(\tau)]^{1/2}d\tau$ is a decreasing function of α. Therefore, $V(u_t(\alpha))$ is not constant for α in an interval. This implies $\omega(\varphi)$ is a single orbit and proves the theorem.

It is also possible to analyze the stability and instability properties of the limiting equilibrium point. If c is an equilibrium point of (14.3), let $\beta = g'(c)$. If $x = y + c$ in (14.3), then the linear variational equation for y is given by

$$
\dot{y}(t) = -\int_{-r}^{0} a(-\theta)g'(c)y(t+\theta)d\theta. \tag{14.8}
$$

If $g'(c) > 0$, then we can use the preceding theorem to conclude that every solution of (14.8) approaches zero as $t \to \infty$ if either

(i) $\ddot{a}(s) \not\equiv 0$, or

(ii) if $a(s) = (r-s)/r$ and

$$\frac{m\,2\pi}{(g'(c))^{\frac{1}{2}}} \neq r$$

for all integers m.

If $g'(c) < 0$, then using the negative of the above function $V(\varphi)$ and the instability Theorem 13.3, one sees that the solution $x = 0$ of (14.8) is unstable. Later, we will see that these properties of the linear equation also hold true of the nonlinear equation.

15. AN EQUATION OF VOLTERRA

Consider the system

$$A\ddot{x}(t) + Bx(t) = \int_0^r F(\theta)x(t-\theta)d\theta \tag{15.1}$$

where A,B,F are symmetric $n \times n$ matrices and F is continuously differentiable. Let

$$M = B - \int_0^r F(\theta)d\theta$$

and write (15.1) as

$$\begin{aligned} \dot{x}(t) &= y(t), \\ A\dot{y}(t) &= -Mx(t) + \int_0^r F(\theta)[x(t-\theta) - x(t)]d\theta. \end{aligned} \tag{15.2}$$

Theorem 15.1. (i) If $A > 0$, $M > 0$, $F(\theta) \geq 0$ and there is a θ_0 in $[0,r]$ such that $\dot{F}(\theta_0) < 0$, then every solution of (15.2) approaches zero as $t \to \infty$.

(ii) If $A > 0$, $M > 0$, $\dot{F} \equiv 0$, $F > 0$, then all solutions of (15.2) are bounded and the ω-limit set of any solution of (15.2) must be generated by periodic solutions of period r of the ordinary system

$$\dot{x} = y, \quad A\dot{y} = -Bx. \tag{15.3}$$

(iii) If $A > 0$, $M < 0$, $F(\theta) > 0$, $0 \leq \theta < r$, $\dot{F}(\theta) \leq 0$, $0 \leq \theta < r$ and there is a θ_0 in $[0,r]$ such that $\dot{F}(\theta_0) < 0$, then the solution $x = 0$, $y = 0$ of (15.2) is unstable.

Proof. Let φ,ψ be the initial values for x,y in (15.2) and define

$$V(\varphi,\psi) = \frac{1}{2}\varphi'(0)M\varphi(0) + \frac{1}{2}\psi'(0)A\psi(0)$$

$$+ \frac{1}{2}\int_0^r [\varphi(-\theta) - \varphi(0)]'F(\theta)[\varphi(-\theta) - \varphi(0)]d\theta$$

where $'$ is the transpose, then a few simple calculations yield

$$\dot{V}(\varphi,\psi) = -\frac{1}{2}[\varphi(-r) - \varphi(0)]'F(r)[\varphi(-r) - \varphi(0)]$$

$$(15.4) \qquad + \frac{1}{2}\int_0^r [\varphi(-\theta) - \varphi(0)]'\dot{F}(\theta)[\varphi(-\theta) - \varphi(0)]d\theta \leq 0$$

if the conditions of (i), (ii) or (iii) are satisfied. We now apply the theorems of Section 13. If either conditions (i) or (iii) are satisfied, then there is an interval I_{θ_0} containing θ_0 such that $\dot{F}(\theta) < 0$ for θ in I_{θ_0}. From (15.4), $\dot{V}(\varphi,\psi) = 0$ implies $\varphi(-\theta) = \varphi(0)$ for $\theta \in I_{\theta_0}$. For a solution x,y to belong to to the largest invariant set where $\dot{V}(\varphi,\psi) = 0$, we must, therefore, have $x(t-\theta) = x(t)$ for all t in $(-\infty,\infty)$, θ in I_{θ_0}. Therefore, $x(t) =$ constant. From equation (15.2), this implies $y = 0$ and thus $Mx = 0$. But, $M > 0$ implies $x = 0$. Therefore, the largest invariant set in the set where $\dot{V}(\varphi,\psi) = 0$ is $\{(0,0)\}$, the origin. If the conditions in (i) are satisfied, then V satisfies Theorem 13.1 and every solution of (15.2) approaches $(0,0)$ as $t \to \infty$. If the conditions in (iii) are satisfied, then V satisfies Theorem 13.2 and $(0,0)$ is unstable.

If the conditions of (ii) are satisfied, then $\dot{F}(\theta) \equiv 0$ and $\dot{V}(\varphi,\psi) = 0$ implies $\varphi(-r) = \varphi(0)$. Thus the largest invariant set in the set where $\dot{V}(\varphi,\psi) = 0$ consists of the r-periodic solutions of equation (15.2). On the other hand, if $x(t),y(t)$ is an r-periodic solution of (15.2) with $F(\theta) =$ constant, then the fact that $M > 0$ implies the integrals of x,y over the interval $[0,r]$ are zero. Therefore, $x(t),y(t)$ must satisfy (15.3). This proves the theorem.

16. NONHOMOGENEOUS LINEAR SYSTEMS

In this section, we consider the linear system

$$(16.1)\qquad \begin{cases} \dot{x}(t) = L(t,x_t) + h(t), & t \geq \sigma \\ x_\sigma = \varphi \end{cases},$$

or equivalently

$$(16.2)\qquad \begin{cases} x(t) = \varphi(0) + \int_\sigma^t L(s,x_s)ds + \int_\sigma^t h(s)ds, & t \geq \sigma \\ x_\sigma = \varphi \end{cases},$$

where $h \in \mathcal{L}_1^{loc}([\sigma,\infty),R^n)$, the space of functions mapping $[\sigma,\infty) \to R^n$ which are Lebesgue integrable on each compact set of $[\sigma,\infty)$. Also, we assume $L(t,\varphi)$ is linear in φ and, in addition, there is an $n \times n$ matrix function $\eta(t,\theta)$ measurable in t,θ, of bounded variation in θ on $[-r,0]$ for each t, and there is an $\ell \in \mathcal{L}_1^{loc}((-\infty,\infty),R)$ such that

$$(16.3)\qquad L(t,\varphi) = \int_{-r}^0 [d_\theta \eta(t,\theta)]\varphi(\theta)$$

$$(16.4)\qquad |L(t,\varphi)| \leq \ell(t)|\varphi|$$

for all $t \in (-\infty,\infty)$, $\varphi \in C$.

The most general type of linear systems with finite lag known to be useful in the applications have the form

$$(16.5)\qquad \begin{aligned} \dot{x}(t) &= \sum_{k=1}^N A_k x(t-\omega_k) \\ &\quad + \int_{-r}^0 A(t,\theta)x(t+\theta)d\theta + h(t) \qquad 0 \leq \omega_k \leq r \end{aligned}$$

where $A(t,\theta)$ is integrable in θ for each t and there is a function $a(t) \in \mathcal{L}_1^{loc}((-\infty,\infty),R)$ such that

$$(16.6) \qquad \left|\int_{-r}^{0} A(t,\theta)\varphi(\theta)d\theta\right| \leq a(t)|\varphi|$$

for all $t \in R$, $\varphi \in C$.

Theorem 16.1. Suppose the above hypotheses on L are satisfied. For any given $\sigma \in R$, $\varphi \in C([-r,0],R^n)$, $h \in \mathcal{L}_1^{loc}([\sigma,\infty),R^n)$, there exists a unique function $x(\sigma,\varphi)$ defined and continuous on $[\sigma-r,\infty)$ that satisfies (16.1) on $[\sigma,\infty)$.

Proof. The condition (16.4) on $L(t,\varphi)$ implies the Caratheodory conditions are satisfied by $L(t,\varphi) + h(t)$. Therefore, we have local existence. Local uniqueness is also a consequence of our general theorem. To prove global existence, we obtain estimates on the solutions which will also be useful later. Let x be a non-continuable solution of (16.1) on $[\sigma-r,b)$. From (16.2)

$$|x(t)| \leq |\varphi(0)| + \int_{\sigma}^{t} \ell(s)|x_s|ds + \left|\int_{\sigma}^{t} h(s)ds\right|$$

$$|x_\sigma| = |\varphi|$$

for all values of $t \in [\sigma,b)$. Thus,

$$|x_t| \leq |\varphi| + \int_{\sigma}^{t} \ell(s)|x_s|ds + \left|\int_{\sigma}^{t} h(s)ds\right|$$

for $t \in [\sigma,b)$. Gronwall's inequality implies

$$(16.7) \qquad \begin{aligned} |x_t| &\leq \left[|\varphi| + \left|\int_{\sigma}^{t} h(s)ds\right|\right]e^{\int_{\sigma}^{t} \ell(s)ds} \\ &\leq \left[|\varphi| + \int_{\sigma}^{t} |h(s)|ds\right]e^{\int_{\sigma}^{t} \ell(s)ds} \end{aligned}$$

for $t \in [\sigma,b)$. But, this relation and the continuation theorem clearly imply the solution is defined on $[\sigma,\infty)$. In fact, if $b \neq \infty$, then the solution must leave every bounded set as $t \to b$, but this contradicts (16.7) and proves the theorem.

Let $x(\sigma,\varphi,h)$ be the solution of (16.1) with initial value φ at σ. Then linearity of L and uniqueness of solutions of (16.1) implies

$$x(\sigma,\varphi,h) = x(\sigma,\varphi,0) + x(\sigma,0,h) \tag{16.8}$$

and relation (16.7) implies

$$\begin{aligned} &(a)\quad |x(\sigma,\varphi,0)(t)| \leq |\varphi| e^{\int_\sigma^t \ell(s)ds}, \quad t \geq \sigma, \\ &(b)\quad |x(\sigma,0,h)(t)| \leq \left(\int_\sigma^t |h(s)|ds\right) e^{\int_\sigma^t \ell(s)ds}, \quad t \geq \sigma. \end{aligned} \tag{16.9}$$

Also, $x(\sigma,\varphi,0)$ is linear in φ and $x(\sigma,0,h)$ is linear in h. Relation (16.9) implies that for each t in $[\sigma,\infty)$ the function $x(\sigma,\cdot,0)(t)\colon C \to R^n$ is a continuous linear map and the function $x(\sigma,0,\cdot)(t)\colon \mathcal{L}_1([0,t],R^n) \to R^n$ is a continuous linear map.

Let us consider $x(\sigma,0,\cdot)(t)$ in detail. We need the following well known theorem from functional analysis.

Theorem 16.2. Suppose $T\colon \mathcal{L}_1([a,b],R^n) \to R^n$ is a continuous linear operator. Then there exists a unique $n \times n$ matrix function $V(\theta)$, $a \leq \theta \leq b$, (unique except for sets of measure zero in θ) which is integrable and essentially bounded such that

$$Th = \int_a^b V(\theta)h(\theta)d\theta, \quad h \in \mathcal{L}_1([a,b],R^n).$$

Theorem 16.3. (variation of constants). If $h \in \mathcal{L}_1^{loc}([\sigma,\infty),R^n)$, L satisfies the above hypothesis and $x(\sigma,\varphi,h)$ is the solution of (16.1) then

$$(16.10) \qquad x(\sigma,\varphi,h)(t) = x(\sigma,\varphi,0)(t) + \int_\sigma^t U(t,s)h(s)ds, \quad t \geq \sigma,$$

$$x_\sigma = \varphi,$$

where $U(t,s)$ is the solution of the equation

$$(16.11) \qquad \begin{cases} U(t,s) = \int_s^t L(u,U_u(\cdot,s))du + I & \text{a.e. in } s \text{ for } t \geq s \\ U(t,s) = 0, & s - r \leq t < s \end{cases}$$

or

$$\begin{cases} \dfrac{\partial U(t,s)}{\partial t} = L(t,U_t(\cdot,s)), \quad t \geq s, \text{ a.e. in } s \text{ and } t \\ U(t,s) = \begin{cases} 0 & s - r \leq t < s \\ I & t = s \end{cases} \end{cases}$$

where $U_t(\cdot,s)(\theta) = U(t+\theta,s)$, $-r \leq \theta \leq 0$.

<u>Proof</u>. From the representation Theorem 16.2, there exists an $n \times n$ matrix function $U^*(t,\sigma,\cdot) \in \mathcal{L}_\infty([\sigma,t],R^{n^2})$, $t \geq \sigma$, such that

$$x(\sigma,0,h)(t) = \int_\sigma^t U^*(t,\sigma,s)h(s)ds, \quad t \geq \sigma.$$

We now show that $U^*(t,\sigma,s)$ is independent of σ. Let α be in $[\sigma,t]$ and let k be any element of $\mathcal{L}_1([\sigma,t],R^n)$ that satisfies $k(s) = 0$ for $s \in [\sigma,\alpha]$. Then $x(\sigma,0,k)(t) = x(\alpha,0,k)(t)$ for $t \geq \alpha$ since $x(\sigma,0,k)(t) = 0$ for $\sigma - r \leq t \leq \alpha$. This implies

$$\int_\alpha^t [U^*(t,\sigma,s) - U^*(t,\alpha,s)]k(s)ds = 0$$

for all $k \in \mathcal{L}_1([\alpha,t],R^n)$. Thus, $U^*(t,\sigma,s) = U^*(t,\alpha,s)$ a.e. in s. Since α is an arbitrary element of $[\sigma,t]$, it follows that $U^*(t,\sigma,s)$ is independent of σ. Define $U(t,s) = U^*(t,\sigma,s)$ for $s \leq t$, $U(t,s) = 0$ for $s - r \leq t < s$.

From (16.2), we have

$$\begin{aligned}
\int_\sigma^t U(t,s)h(s)ds
&= \int_\sigma^t \left(\int_{-r}^0 [d_\theta \eta(s,\theta)] \int_\sigma^{s+\theta} U(s+\theta,u)h(u)du\right)ds + \int_\sigma^t h(s)ds \\
&= \int_\sigma^t \left(\int_{-r}^0 [d_\theta \eta(s,\theta)] \int_\sigma^{s} U(s+\theta,u)h(u)du\right)ds + \int_\sigma^t h(s)ds \\
&= \int_\sigma^t \left[\int_u^t \left(\int_{-r}^0 [d_\theta \eta(s,\theta)] U(s+\theta,u)\right)ds\right]h(u)du \\
&\qquad + \int_\sigma^t h(s)ds \\
&= \int_\sigma^t \left(\int_s^t \left\{\int_{-r}^0 d_\theta \eta(u,\theta) U(u+\theta,s)\right\}du\right)h(s)ds \\
&\qquad + \int_\sigma^t h(s)ds
\end{aligned}$$

for all $h \in \mathcal{L}_1([\sigma,t],R^n)$. Therefore

$$U(t,s) = \int_s^t \left(\int_{-r}^0 [d_\theta \eta(u,\theta)] U(u+\theta,s)\right)du + I \quad \text{a.e.}$$

If we differentiate with respect to t, we have

$$(16.12) \qquad \begin{aligned}
\frac{\partial U(t,s)}{\partial t} &= \int_{-r}^0 [d_\theta \eta(t,\theta)] U(t+\theta,s) = L(t,U_t(\cdot,s)), \quad t \geq s \\
U(t,s) &= \begin{cases} 0 & \text{for } s - r \leq t < s \\ I & \text{for } t = s \end{cases}
\end{aligned}$$

and $U_t(\cdot,s)(\theta) = U(t+\theta,s)$, $-r \leq \theta \leq 0$. This proves the theorem.

<u>Corollary 16.1.</u> If $L(t,\varphi) \equiv L(\varphi)$ is independent of t in (16.1), then $U(t,s) = U(t-s,0) \overset{\text{def}}{=} U(t-s)$ and

$$(16.13) \qquad \begin{cases} x(\sigma,\varphi,h)(t) = x(t-\sigma,\varphi,0)(0) + \int_\sigma^t U(t-s)h(s)ds, & t \geq \sigma \\ x_\sigma = \varphi \end{cases}$$

Proof. This is clear from (16.11).

It is convenient to have the variation of constants formula written in a different manner. From (16.10), we have

$$x(\sigma,\varphi,h)(t+\theta) = x(\sigma,\varphi,0)(t+\theta) + \int_\sigma^{t+\theta} U(t+\theta,s)h(s)ds, \quad t+\theta \geq \sigma$$

$$x(\sigma,\varphi,h)(t+\theta) = \varphi(t+\theta), \quad \sigma - r \leq t+\theta \leq \sigma, \quad -r \leq \theta \leq 0.$$

Since $U(t+\theta,s) = 0$ for $s > t+\theta$, and $x(\sigma,\varphi,0)(t+\theta) = \varphi(t-\sigma+\theta)$ for $\sigma - r \leq t+\theta \leq \sigma$, both of these expressions can be combined into the single expression

$$x(\sigma,\varphi,h)(t+\theta) = x(\sigma,\varphi,0)(t+\theta) + \int_\sigma^t U(t+\theta,s)h(s)ds$$

$$t \geq \sigma, \quad -r \leq \theta \leq 0,$$

or

$$x_t(\sigma,\varphi,h)(\theta) = x_t(\sigma,\varphi,0)(\theta) + \int_\sigma^t U_t(\cdot,s)(\theta)h(s)ds,$$

$$t \geq \sigma, \quad -r \leq \theta \leq 0,$$

or

$$(16.14) \qquad x_t(\sigma,\varphi,h) = x_t(\sigma,\varphi,0) + \int_\sigma^t U_t(\cdot,s)h(s)ds, \quad t \geq \sigma,$$

where it is always understood that the integral equation (16.14) is actually an integral in Euclidean space. All of the usual operations for integrals are valid for (16.14).

If we introduce some more notation, the integral equation becomes even nicer. If

$$x_t(\sigma,\varphi,0) \overset{\text{def}}{=} T(t,\sigma)\varphi, \tag{16.15}$$

then $T(t,\sigma)$ is a continuous linear operator. Furthermore, since $U(t,s)$ is continuous in t for $t > s$, it follows from (16.11) that $U(t,s)$ has a continuous first derivative in t for $t > s + r$. Thus, (16.12) is satisfied exactly in t for all $t \geq s + r$ and a.e. in s. Therefore, we are justified in writing

$$\begin{gathered} U_t(\cdot,s) = T(t,s)X_0 \\ X_0(\theta) = \begin{cases} 0 & -r \leq \theta < 0 \\ I & \theta = 0 \end{cases} \end{gathered} \tag{16.16}$$

With this notation, the integral equation becomes

$$x_t(\sigma,\varphi,h) = T(t,\sigma)\varphi + \int_\sigma^t T(t,s)X_0 h(s)ds, \quad t \geq \sigma. \tag{16.17}$$

This is the formula we use very often and it has great advantages over the original variation of constants formula.

The first advantage arises in the following manner. In the equation (16.1) the operation of introducing a new variable $x_t = e^{\alpha t} z_t$ is not valid. In fact, for this to be a valid operation (i.e. for z_t to satisfy a functional differential equation), we must have $z_t(\theta) = z(t+\theta)$ for all $t \in (-\infty,\infty)$, $\theta \in [-r,0]$. This is clearly not satisfied. On the other hand, in the integral equation (16.17) we can make such a transformation, discuss the behavior of z as a solution of the integral equation in spite of the fact that it does not satisfy a functional differential equation and then return to x_t to obtain information about the original equation.

A second advantage which will become more apparent later is the following. If C is decomposed as a direct sum of two subspaces, we can make changes of variables in a subspace and determine the integral equations for the new variables in a subspace. In particular, for constant coefficients, this is analogous to using the Jordan canonical form.

17. THE "ADJOINT" EQUATION AND REPRESENTATION OF SOLUTIONS

In this section, we restrict our attention to the linear system

$$\dot{x}(t) = L(t,x_t) \tag{17.1}$$

where $L(t,\varphi)$ is continuous in t,φ, linear in φ and is given explicitly by

$$L(t,\varphi) = \sum_{k=1}^{\infty} A_k(t)\varphi(-\tau_k) + \int_{-\tau}^{0} A(t,\xi)\varphi(\xi)d\xi \tag{17.2}$$

where each $A_k(t)$, $A(t,\xi)$ are continuous $n \times n$ matrix functions for $-\infty < t,\xi < \infty$, $0 \leq \tau_k$, $\tau \leq r$. The extension of the results of this section to the most general linear system is contained in Section 32.

We define the equation "adjoint" to (17.1) as

$$\frac{dy(s)}{ds} = -\sum_{k=1}^{\infty} y(s+\tau_k)A_k(s+\tau_k) - \int_{-\tau}^{0} y(s-\xi)A(s-\xi,\xi)d\xi. \tag{17.3}$$

Let $C^* = C([0,r],R^{n*})$ be the space of continuous functions mapping $[0,r]$ into the n-dimensional row vectors and for any $\psi \in C^*$, $\varphi \in C$, $t \in R$, let

$$(\psi,\varphi,t) = \psi(0)\varphi(0) - \sum_{k=1}^{\infty}\int_{0}^{\tau_k}\psi(\xi)A_k(t+\xi)\varphi(\xi-\tau_k)d\xi - \int_{-\tau}^{0}\left(\int_{-\theta}^{0}\psi(\xi)A(t+\xi,\theta)\varphi(\xi+\theta)d\xi\right)d\theta. \tag{17.4}$$

For any $\psi \in C^*$, let $y(\sigma,\psi)$ be the solution of (17.3) on $(-\infty,\sigma+r]$ with $y(\sigma,\psi)(\sigma+s) = \psi(s)$, $0 \leq s \leq r$. Also, let $y^t(\sigma,\psi) \in C^*$, $t \leq \sigma$, be defined by $y^t(\sigma,\psi)(s) = y(\sigma,\psi)(t+s)$, $0 \leq s \leq r$.

If x is a solution of (17.1) on $[\sigma-r,T]$ and y is a solution of

(17.3) on $[\sigma,T+r]$, then, for $\sigma \le t \le T$,

$$
(17.5)\qquad
\begin{aligned}
(y^t,x_t,t) &= y(t)x(t) - \sum_{k=1}^{\infty}\int_t^{t+\tau_k} y(\alpha)A_k(\alpha)x(\alpha-\tau_k)d\alpha \\
&\quad -\int_{-\tau}^{0}\Big[\int_{t-\theta}^{t} y(\alpha)A(\alpha,\theta)x(\alpha+\theta)d\alpha\Big]d\theta.
\end{aligned}
$$

A few straightforward calculations using (17.1), (17.3) and (17.5) yields $\frac{d}{dt}(y^t,x_t,t) = 0$, $\sigma \le t \le T$, and, therefore, we have proved

<u>Lemma 17.1.</u> If x is a solution of (17.1) on $[\sigma-r,T]$ and y is a solution of (17.3) on $[\sigma,T+r]$, then $(y^t,x_t,t) = \text{constant}$ for $\sigma \le t \le T$.

Suppose $U(t,s)$ is the $n \times n$ matrix solution of (17.1) on $[s-r,\infty)$ defined at $t = s$ by the initial $n \times n$ matrix $U(t,s) = 0$, $s-r \le t < s$, $U(s,s) = I$, the identity. The matrix $U(t,s)$ satisfies (16.11). The bilinear form (ψ,φ,t) obviously is meaningful for piecewise continuous functions. As in the proof of Lemma 17.1, one shows that if y is any solution of (17.3) on $[s,t+r]$, then $(y^\sigma,U_\sigma(\cdot,s),\sigma) = \text{constant}$ for $s \le \sigma \le t$, where $U_\sigma(\cdot,s)(\theta) = U(\sigma+\theta,s)$, $-r \le \theta \le 0$. In particular, $(y^s,U_s(\cdot,s),s) = (y^t,U_t(\cdot,s),t)$. From the special nature of $U_s(\cdot,s)$, it follows that

$$
(17.6)\qquad y(s) = (y^t,U_t(\cdot,s),t), \quad s \le t.
$$

In a similar manner, if $V(s,t)$ is the $n \times n$ matrix solution of (17.3) on $(-\infty,t+r]$ defined at $s = t$ by the initial $n \times n$ matrix $V(s,t) = 0$, $t < s \le t+r$, $V(t,t) = I$, the identity, then

$$
(17.7)\qquad x(t) = (V^s(\cdot,t),x_s,s), \quad t \ge s,
$$

where $V^s(\cdot,t)(\alpha) = V(s+\alpha,t)$, $0 \le \alpha \le r$. If we apply formula (17.7) to the matrix $V(t,s)$, then we obtain

$$(17.8) \qquad U(t,s) = V(s,t), \quad t \geq s,$$

and, thus, the solution of the homogeneous equation (17.1) is representable by the matrix solution $U(t,s)$ of (17.1) and the bilinear form (17.4). If we combine these remarks with the variation of constants formula, we have

<u>Theorem 17.1.</u> If $V(s,t)$ is the $n \times n$ matrix solution of the adjoint equation (17.3) on $(-\infty, t+r]$, $V(s,t) = 0$, $t < s \leq t+r$, $V(t,t) = I$, and $f \in C((-\infty,\infty),R^n)$, then any solution x of the equation

$$(17.9) \qquad \dot{x}(t) = L(t,x_t) + f(t)$$

defined for $t \geq s$ satisfies the relation

$$(17.10) \qquad x(t) = (V^s(\cdot,t),x_s,s) + \int_s^t V(\alpha,t)f(\alpha)d\alpha, \quad t \geq s.$$

The relationship of the solution operator of the adjoint equation (17.3) to the functional analytical adjoint of the solution operator of (17.1) will be discussed in Section 33.

18. STABILITY OF PERTURBED SYSTEMS

In this section, we suppose that L satisfies the conditions of Section 17; namely, that the representation (17.2) holds. Furthermore, if

$$(18.1) \qquad |L(t,\varphi)| \leq \ell(t)|\varphi|, \quad t \in R^+, \varphi \in C,$$

then there is a constant ℓ_1 such that

$$(18.2) \qquad \int_t^{t+r} \ell(s)ds \leq \ell_1, \quad t \in R^+.$$

We can then prove

Lemma 18.1. If $L(t,\varphi)$ satisfies (18.1) and (18.2) and $U(t,s)$ is defined as in Section 17, then

(i) the solution $x = 0$ of (17.1) is uniformly stable if and only if there is a constant $M > 0$ such that

$$(18.3) \qquad |U(t,s)| \leq M, \quad t \geq s \geq 0$$

(ii) the solution $x = 0$ of (17.1) is uniformly asymptotically stable if and only if there are constants $\alpha > 0$, $M > 0$ such that

$$(18.4) \qquad |U(t,s)| \leq Me^{-\alpha(t-s)}, \quad t \geq s \geq 0.$$

Proof. Let us first estimate $U(t,s)$ for $s \leq t \leq s+r$. Since $U(t,s)$ satisfies (16.11), it follows that

$$|U(t,s)| \leq \int_s^t \ell(u)|U_u(\cdot,s)|du + 1, \quad t \geq s$$
$$|U_s(\cdot,s)| = 1.$$

Therefore,

$$|U_t(\cdot,s)| \le 1 + \int_s^t \ell(u)|U_u(\cdot,s)|du, \quad t \ge s$$

and Gronwall's inequality implies

$$|U_t(\cdot,s)| \le e^{\int_s^t \ell(u)du}, \quad t \ge s.$$

From the hypothesis on $\ell(u)$, this implies

$$|U(t,s)| \le e^{\ell_1}, \quad s \le t \le s+r. \tag{18.5}$$

Case (i). As in ordinary differential equations, the solution $x = 0$ of (17.1) is uniformly stable if and only if there is an $M_1 > 0$ such that $|x_t(\sigma,\varphi)| \le M_1|\varphi|$, $t \ge \sigma \ge 0$, $\varphi \in C$. If system (17.1) is uniformly stable, the fact that $U(t,s)$ satisfies (17.1) for $t \ge s+r$, implies $|U(t,s)| \le M_1$, $t \ge s+r$. If $M = \max(M_1, \exp \ell_1)$, then (18.5) implies (18.3). Conversely, if (18.3) holds, then (17.8) and (17.10) imply the existence of an M_1 such that $|x_t(\sigma,\varphi)| \le M_1|\varphi|$ for $t \ge \sigma$, $\varphi \in C$. This proves (i).

Case (ii). As in ordinary differential equations, uniform asymptotic stability of the solution $x = 0$ of (17.1) is equivalent to the existence of constants $\alpha > 0$, $M_1 > 0$ such that $|x_t(\sigma,\varphi)| \le M_1|\varphi| \exp[-\alpha(t-\sigma)]$, $t \ge \sigma \ge 0$, $\varphi \in C$. The remainder of the proof is similar to Case (i).

Theorem 18.1. Suppose system (17.1) is uniformly stable. If $g: R^+ \times C \to R^n$ is continuous and there is a $\gamma \in \mathcal{L}_1(R^+,R)$ such that

$$|g(t,\varphi)| \le \gamma(t)|\varphi|, \quad t \in R^+, \ \varphi \in C, \tag{18.6}$$

then the system

$$(18.7) \qquad \dot{x}(t) = L(t,x_t) + g(t,x_t)$$

is uniformly stable.

Theorem 18.2. Suppose system (17.1) is uniformly asymptotically stable. If g satisfies (18.6) and there are constants $\tau > 0$, $\beta > 0$ such that

$$\int_\sigma^t \gamma(s)ds \le \beta(t-\sigma) + \tau, \quad t \ge \sigma \ge 0,$$

then there is a constant $\delta > 0$ such that system (18.7) is uniformly asymptotically stable if $\gamma < \delta$.

Theorem 18.3. Suppose system (17.1) is uniformly asymptotically stable. If $f: R^+ \times C \to R^n$ is continuous and for every $\varepsilon > 0$, there is a $\delta > 0$ such that

$$|f(t,\varphi)| \le \varepsilon|\varphi| \quad \text{for} \quad t \in R^+, \ \varphi \in C, \ |\varphi| < \delta,$$

then the solution $x = 0$ of

$$\dot{x}(t) = L(t,x_t) + f(t,x_t)$$

is uniformly asymptotically stable.

We do not give the proofs of these theorems since they are exactly the same as the analogous ones for ordinary differential equations if one uses (17.8), (17.10) and Lemma 18.1.

19. LINEAR AUTONOMOUS EQUATIONS. THE SEMIGROUP AND INFINITESIMAL GENERATOR

We are now going to restrict our discussion to a special class of linear equations in order to obtain a better understanding of the geometry of solutions. More specifically, we consider linear autonomous functional differential equations by which we understand equations of the form

$$\dot{x}(t) = L(x_t) \tag{19.1}$$

where L is a continuous linear function mapping C into R^n. There exists an $n \times n$ matrix $\eta(\theta)$, $-r \leq \theta \leq 0$, whose elements are of bounded variation such that

$$L(\varphi) = \int_{-r}^{0} [d\eta(\theta)]\varphi(\theta), \quad \varphi \text{ in } C. \tag{19.2}$$

The goal is to understand the geometric behavior of the solutions of (19.1) when they are interpreted in C. More specifically, we shall see that the behavior of the solutions of (19.1) on an eigenspace can be interpreted as a solution of an ordinary differential equation. General results of functional analysis imply the existence of a coordinate system in C which exhibits the eigenspace as well as a complementary subspace which is invariant under the transformation induced by the solutions of (19.1). An explicit characterization of this coordinate system will be given with the aid of the equation adjoint to (19.1).

If φ is any given function in C and $x(\varphi)$ is the unique solution of (19.1) with the initial function φ at zero, we define the operator $T(t)$ mapping C into C by the relation

$$x_t(\varphi) = T(t)\varphi. \tag{19.3}$$

Lemma 19.1. The operator $T(t)$, $t \geq 0$, defined on C by (19.3) satisfies the following relations

(i) The family $\{T(t), t \geq 0\}$ is a semigroup of linear transformations;

that is $T(t+\tau) = T(t)T(\tau)$ for all $t \geq 0$, $\tau \geq 0$.

(ii) $T(t)$ is bounded for each $t \geq 0$, $T(0) = I$, and $T(t)$ is strongly continuous on $[0,\infty)$; that is,

$$\lim_{\tau \to t} |T(t)\varphi - T(\tau)\varphi| = 0$$

for all $t \geq 0$, φ in C.

(iii) $T(t)$ is completely continuous (compact) for $t \geq r$; that is, $T(t)$, $t \geq r$, is continuous and maps bounded sets into relatively compact sets.

Proof. (i) From the uniqueness of solutions of (19.1), it is obvious that $T(t)$ is a linear transformation. The uniqueness also yields the semigroup property in a straightforward way.

(ii) By definition, $T(0) = I$. Since $L(\varphi)$ is continuous and linear, it follows there is a constant ℓ such that $|L(\varphi)| \leq \ell|\varphi|$ for all φ. From the definition of $T(t)$, we have, for any fixed t,

$$
\begin{aligned}
&T(t)\varphi(\theta) = \varphi(t+\theta), \quad t+\theta \leq 0, \\
(19.4) \qquad &T(t)\varphi(\theta) = \varphi(0) + \int_0^{t+\theta} L(T(s)\varphi)ds, \quad t+\theta > 0, \\
&-r \leq \theta \leq 0.
\end{aligned}
$$

It follows that $|T(t)\varphi| \leq |\varphi| + \int_0^t \ell|T(s)\varphi|ds$. Gronwall's inequality then implies that

$$(19.5) \qquad |T(t)\varphi| \leq e^{\ell t}|\varphi|, \quad t \geq 0, \ \varphi \text{ in } C,$$

and, thus, $T(t)$ is bounded.

Since $T(0) = I$, $T(t)$ is bounded and $T(t)$ satisfies the semi-group property it is readily seen that $T(t)$ will be strongly continuous on $[0,\infty)$ if $\lim_{t \to 0^+} |T(t)\varphi - \varphi| = 0$. But, this latter property is obvious from (19.4).

(iii) If $S = \{\varphi$ in $C: |\varphi| \leq R\}$, then for any ψ in $T(t)S$, $t \geq r$, relation (19.5) implies $|\psi| \leq e^{\ell t}R$, and (19.1) implies $|\dot{\psi}| \leq \ell e^{\ell t}R$. Since these functions are uniformly bounded with a uniform lipschitz constant, $T(t)S$, $t \geq r$, belongs to a compact subset of C. This completes the proof of Lemma 19.1.

Since $T(t)$ is strongly continuous we may define the infinitesimal operator A of $T(t)$ [Hille and Phillips, p. 306] as

$$A\varphi = \lim_{t \to 0^+} \frac{1}{t}[T(t)\varphi - \varphi] \tag{19.6}$$

whenever this limit exists. Of course, the limit signifies convergence in the norm in C. The infinitesimal generator of $T(t)$ is the smallest closed extension of A. An operator is closed if φ_n in $\mathscr{D}(A)$, the domain of A, $\varphi_n \to \varphi$ and $\psi_n = A\varphi_n \to \psi$ implies φ is in $\mathscr{D}(A)$ and $\psi = A\varphi$. If $T(0) = I$ and $\lim_{t \to 0^+} T(t)\varphi = \varphi$ for all φ in C, then it follows from a Corollary, p. 344 and Theorem 10.6.1, p. 322 of Hille and Phillips that the infinitesimal operator of $T(t)$ is closed and, thus, the infinitesimal generator of $T(t)$ is given by (19.6). From the above remarks and Theorem 10.3.1 of Hille and Phillips, p. 307, the domain of A, $\mathscr{D}(A)$, is dense in C and the range of A, $\mathscr{R}(A)$, is in C. This allows one to compute the infinitesimal generator directly from (19.6) and (19.4). For any θ in $[-r,0)$, it follows directly from the first relation in (19.4) that

$$\lim_{t \to 0^+} \frac{1}{t}[T(t)\varphi(\theta) - \varphi(\theta)] = \frac{d\varphi(\theta+)}{d\theta}$$

where $d\varphi(\theta+)/d\theta$ is the right hand derivative at θ. If $\theta = 0$, then the second relation in (19.4) yields

$$\begin{aligned} \lim_{t \to 0^+} \frac{1}{t}[T(t)\varphi(0) - \varphi(0)] &= \lim_{t \to 0^+} \frac{1}{t}\int_0^t L(T(s)\varphi)ds \\ &= L(T(0)\varphi) = L(\varphi) = \int_{-r}^0 [d\eta(\theta)]\varphi(\theta). \end{aligned}$$

Since $\mathscr{R}(A)$ is in C, it follows from the definition of $A\varphi$ on $[-r,0)$ that φ is in $\mathscr{D}(A)$ if and only if φ is continuous and has a continuous right hand derivative in $[-r,0)$, which implies φ has a continuous derivative in $[-r,0)$. But, in addition $A\varphi(0) = L(\varphi)$ and $A\varphi$ continuous implies φ is in $\mathscr{D}(A)$ if and only if φ has a continuous derivative on $[-r,0]$ and $\dot{\varphi}(0) = L(\varphi)$. Finally, $\mathscr{R}(A)$ is given by

$$(19.7) \qquad A\varphi(\theta) = \begin{cases} d\varphi(\theta)/d\theta, & -r \leq \theta < 0 \\ \\ L(\varphi) \stackrel{\text{def}}{=} \int_{-r}^{0} [d\eta(\theta)]\varphi(\theta), & \theta = 0 \end{cases}$$

Summarizing these results and using Theorem 10.3.3 of Hille and Phillips, p. 308, we have the following

Lemma 19.2. The infinitesimal generator A of the family of transformations $T(t)$, $t \geq 0$ defined by (19.3) is given by (19.7), $\mathscr{D}(A)$ is dense in C, and for any φ in $\mathscr{D}(A)$,

$$\frac{d}{dt} T(t)\varphi = T(t)A\varphi = AT(t)\varphi.$$

20. THE EIGENVALUES OF A LINEAR AUTONOMOUS EQUATION. DECOMPOSITION OF C.

To proceed further, we need the definition of the spectrum of a linear operator B taking a Banach space $\mathscr{B}$ into itself. The <u>resolvent set</u> $\rho(B)$ of B is the set of values λ in the complex plane for which the operator $\lambda I-B$, I the identity operator, has a bounded inverse with domain dense in $\mathscr{B}$. The complement of $\rho(B)$ in the complex plane is called the <u>spectrum</u> of B and is denoted by $\sigma(B)$. The spectrum $\sigma(B)$ of an operator may consist of three different types of points, namely the <u>residual spectrum</u> $R\sigma(B)$, the <u>continuous spectrum</u> $C\sigma(B)$ and the <u>point spectrum</u> $P\sigma(B)$. The residual spectrum consists of those λ in $\sigma(B)$ for which $(\lambda I-B)^{-1}$ exists but $\mathscr{D}(\lambda I-B)^{-1}$ is not dense in $\mathscr{B}$. The continuous spectrum consists of those λ in $\sigma(B)$ for which $\lambda I-B$ has an unbounded inverse with dense domain the point spectrum consists of those values of λ in $\sigma(B)$ for which $\lambda I-B$ does not have an inverse. The points λ in $P\sigma(B)$ are sometimes called the <u>eigenvalues</u> of B and any nonzero φ in B such that $(\lambda I-B)\varphi = 0$ is called an <u>eigenvector</u>. The <u>null space</u> $\mathfrak{N}(B)$ of B is the set of all φ in $\mathscr{B}$ for which $B\varphi = 0$. For a given λ in $\sigma(B)$ the <u>generalized eigenspace</u> of λ will be denoted by $\mathfrak{M}_\lambda(B)$ and is defined to be the smallest subspace of $\mathscr{B}$ containing all the elements of $\mathscr{B}$ which belong to $\mathfrak{N}(\lambda I-B)^k$, $k = 1,2,\ldots$.

One of our first objectives is to try to determine the nature of $\sigma(T(t))$ and $\sigma(A)$ for the operators which arise in (19.3) and (19.7) and to analyze in what sense the operator $T(t)$ is approximated by e^{At} provided this latter object makes sense. For the simple case in which system (19.1) is an ordinary differential equation; that is, $r = 0$, then $L(\varphi) = A\varphi(0)$ for a constant matrix A, the operator $T(t)$ is e^{At} and A is the infinitesimal generator of $T(t)$. We shall see below that the situation for the more general problem (19.1) is very analogous. Since $T(t)$ will generally not be known, we would hope to discuss most of the properties of $T(t)$ by using only properties of the known operator A in (19.7). Therefore, we first prove

Lemma 20.1. If A is defined by (19.7), then $\sigma(A) = P\sigma(A)$ and λ is in $\sigma(A)$ if and only if λ satisfies the characteristic equation

$$\det \Delta(\lambda) = 0, \quad \Delta(\lambda) = \lambda I - \int_{-r}^{0} e^{\lambda\theta} d\eta(\theta). \tag{20.1}$$

The roots of (20.1) have real parts bounded above and for any λ in $\sigma(A)$ the generalized eigenspace $\mathfrak{M}_\lambda(A)$ is finite dimensional. Finally, there is an integer k such that $\mathfrak{M}_\lambda(A) = \mathfrak{N}(A-\lambda I)^k$ and

$$C = \mathfrak{N}(A-\lambda I)^k \oplus \mathscr{R}(A-\lambda I)^k$$

where the symbol $\oplus$ of course means the direct sum.

Proof. To prove that $\sigma(A) = P\sigma(A)$, we show the resolvent set $\rho(A)$ consists of all λ except those which satisfy (20.1) and then show that any λ satisfying (20.1) is in $P\sigma(A)$. The constant λ will be in $\rho(A)$ if and only if the equation

$$(A-\lambda I)\varphi = \psi \tag{20.2}$$

has a solution φ in $\mathscr{D}(A)$ for every ψ in a dense set in C and the solution depends continuously upon ψ. We will actually solve (20.2) for any ψ in C.

Since any φ in $\mathscr{D}(A)$ must be continuously differentiable and $A\varphi(\theta) = \dot\varphi(\theta)$, a solution of (20.2) must satisfy

$$\dot\varphi(\theta) - \lambda\varphi(\theta) = \psi(\theta), \quad -r \le \theta \le 0;$$

that is,

$$\varphi(\theta) = e^{\lambda\theta} b + \int_0^\theta e^{\lambda(\theta-\xi)}\psi(\xi)d\xi, \quad -r \le \theta \le 0. \tag{20.3}$$

But, φ will be in $\mathscr{D}(A)$ if and only if $\dot{\varphi} \in C$ and $\dot{\varphi}(0) = L(\varphi) \overset{\text{def}}{=} \int_{-r}^{0} [d\eta(\theta)]\varphi(\theta)$ and this yields

$$\lambda b + \psi(0) = \int_{-r}^{0} [d\eta(\theta)][e^{\lambda\theta} b + \int_{0}^{\theta} e^{\lambda(\theta-\xi)}\psi(\xi)d\xi].$$

Simplifying this expression, we obtain

$$\begin{aligned} \Delta(\lambda)b &= -\psi(0) + \int_{-r}^{0}\int_{0}^{\theta} e^{\lambda(\theta-\xi)} d\eta(\theta)\psi(\xi)d\xi \\ &\overset{\text{def}}{=} -(\alpha,\psi) \end{aligned} \tag{20.4}$$

where α is a continuous matrix function on $[0,r]$ defined by

$$\alpha(s) = e^{-\lambda s}I, \quad 0 \leq s \leq r,$$

and

$$(\alpha,\psi) = \alpha(0)\psi(0) - \int_{-r}^{0}\int_{0}^{\theta} \alpha(\xi-\theta)d\eta(\theta)\psi(\xi)d\xi. \tag{20.5}$$

From, (20.3), (20.4), it is clear that equation (20.2) has a solution for every ψ in C only if $\det \Delta(\lambda) \neq 0$ since the mapping $(\alpha,\cdot)$ covers R^n. Also, $\det \Delta(\lambda) \neq 0$ implies a solution of (20.2) for every ψ in C and this solution depends continuously upon ψ. Therefore $\rho(A) = \{\lambda: \det \Delta(\lambda) \neq 0\}$.

If $\det \Delta(\lambda) = 0$, then (20.3), (20.4) imply there exists a nonzero solution of (20.2) for $\psi = 0$; that is, λ is in $P\sigma(A)$. This proves the first part of the lemma.

The characteristic function $\det \Delta(\lambda)$ is an entire function of λ and therefore has zeros of finite order. This implies from (20.3) and (20.4) that the resolvent operator $(A-\lambda I)^{-1}$ has a pole of order k at λ_0 if λ_0 is a zero of $\det \Delta(\lambda)$ of order k. Since A is a closed operator, it follows from Theorem

5.8-A, p. 306 of Taylor [1] that $\mathfrak{M}_{\lambda_0}(A)$ is finite dimensional and has the properties stated in the lemma.

Since the characteristic function is a polynomial in λ of degree n with leading coefficient one and the lower order terms have coefficients which depend upon λ through integrations over $[-r,0]$ of terms of the form $e^{\lambda\theta}$, it follows immediately that there is a $\gamma > 0$ such that no roots of (20.1) have real parts greater than γ. This completes the proof of Lemma 20.1.

From Lemma 20.1, we know that λ in $\sigma(A)$ implies $\mathfrak{M}_\lambda(A)$ is finite dimensional and $\mathfrak{M}_\lambda(A) = \mathfrak{N}(A-\lambda I)^k$ for some integer k. The subspace $\mathfrak{M}_\lambda(A)$ satisfies $A\mathfrak{M}_\lambda(A) \subset \mathfrak{M}_\lambda(A)$ since φ in $\mathfrak{M}_\lambda(A)$ implies $(A-\lambda I)^k\varphi = 0$ and A commutes with $(A-\lambda I)^k$. Let $\mathfrak{M}_\lambda(A)$ have dimension d, let $\varphi_1^\lambda,\ldots,\varphi_d^\lambda$ be a basis for $\mathfrak{M}_\lambda(A)$, and let $\Phi_\lambda = (\varphi_1^\lambda,\ldots,\varphi_d^\lambda)$. Since $A\mathfrak{M}_\lambda(A) \subset \mathfrak{M}_\lambda(A)$, there is a $d \times d$ constant matrix B_λ such that $A\Phi_\lambda = \Phi_\lambda B_\lambda$. The only eigenvalue of B_λ is λ. In fact, for any d-vector a, $(A-\lambda I)^k\Phi_\lambda a = 0$ and one easily shows this implies $\Phi_\lambda(B_\lambda-\lambda I)^k a = 0$ for all d-vectors a. Therefore, $(B_\lambda-\lambda I)^k a = 0$ for all d-vectors a. But this implies $(B_\lambda-\lambda I)^k = 0$ and the result follows from known results in matrix theory. From the definition of A in (19.7), the relation $A\Phi_\lambda = \Phi_\lambda B_\lambda$ implies that

$$\Phi_\lambda(\theta) = \Phi_\lambda(0)e^{B_\lambda\theta}, \quad -r \leq \theta \leq 0.$$

From Lemma 19.2, one also obtains

$$T(t)\Phi_\lambda = \Phi_\lambda e^{B_\lambda t}$$

for $t \geq 0$, which together with the expression for Φ_λ implies that

$$[T(t)\Phi_\lambda](\theta) = \Phi_\lambda(0)e^{B_\lambda(t+\theta)}, \quad -r \leq \theta \leq 0.$$

This relation permits one to define $T(t)$ on $\mathfrak{M}_\lambda(A)$ for all values of t in

$(-\infty,\infty)$. Therefore, on the generalized eigenspace of an eigenvalue of (19.1); that is, an element of $\sigma(A)$, the differential equation (19.1) has the same structure as an ordinary differential equation.

From Lemma 19.2, we also know that $T(t)A\varphi = AT(t)\varphi$ for all φ in $\mathscr{D}(A)$. This implies that $\mathscr{R}(A-\lambda I)^k$ is also invariant under $T(t)$. By a repeated application of the above process we obtain

Theorem 20.1. Suppose Λ is a finite set $\{\lambda_1,\ldots,\lambda_p\}$ of eigenvalues of (19.1) and let $\Phi_\Lambda = (\Phi_{\lambda_1},\ldots,\Phi_{\lambda_p})$, $B_\Lambda = \operatorname{diag}(B_{\lambda_1},\ldots,B_{\lambda_p})$, where Φ_{λ_j} is a basis for the generalized eigenspace of λ_j and B_{λ_j} is the matrix defined by $A\Phi_{\lambda_j} = \Phi_{\lambda_j}B_{\lambda_j}$, $j = 1,2,\ldots,p$. Then the only eigenvalue of B_{λ_j} is λ_j and for any vector a of the same dimension as Φ_Λ, the solution $T(t)\Phi_\Lambda a$ with initial value $\Phi_\Lambda a$ at $t = 0$ may be defined on $(-\infty,\infty)$ by the relation

$$T(t)\Phi_\Lambda a = \Phi_\Lambda e^{B_\Lambda t} a$$

(20.6)

$$\Phi_\Lambda(\theta) = \Phi_\Lambda(0)e^{B_\Lambda\theta}, \quad -r \le \theta \le 0.$$

Furthermore, there exists a subspace Q_Λ of C such that $T(t)Q_\Lambda \subset Q_\Lambda$ for all $t \ge 0$ and

$$C = P_\Lambda \oplus Q_\Lambda$$

where $P_\Lambda = \{\varphi \text{ in } C\colon \varphi = \Phi_\Lambda a, \text{ for some vector } a\}$.

Theorem 20.1 gives a very clear picture of the behavior of the solutions of (19.1). In fact, on generalized eigenspaces, system (19.1) behaves essentially as an ordinary differential equation and the decomposition of C into two subspaces invariant under A and $T(t)$ tells us that we can separate out the behavior on the eigenspaces from the other type of behavior. The above decomposition of C allows one to introduce a coordinate system in C which plays the same role as the Jordan canonical form in ordinary differential equations. As we know in ordinary differential equations, this is very important for studying systems which are close to linear.

The decomposition of C will be complete provided that we can explicitly characterize the projection operator defined by this decomposition. We shall also need bounds for $T(t)$ on the complementary subspace Q_Λ in order to apply the results to perturbed linear systems. In order not to interrupt the continuity of the ideas, we first give an explicit representation for the projection operator. This will be given in terms of an operator "adjoint" to A relative to a certain bilinear form. The reason for taking this approach is so that the final results can be interpreted in a language familiar to differential equations. Other methods of obtaining this projection operator can be given, but the final result will be the same.

21. DECOMPOSING C WITH THE ADJOINT EQUATION

Before proceeding directly to the formal theory, let us attempt to motivate (as much as we can) the concepts which are going to be introduced. In the proof of Lemma 20.1, namely, formulas (20.4) and (20.5), we encountered a bilinear form in a very natural manner. In fact, if λ is a solution of (20.1), then formulas (20.3), (20.4), (20.5) show that the equation $(A-\lambda I)\varphi = \psi$ has a solution if and only if, for every nonzero row vector a for which $a\Delta(\lambda) = 0$, it follows that $(ae^{-\lambda\cdot}I,\psi) = 0$ where (α,ψ) is defined in (20.5). This certainly suggests an "alternative" theorem and indicates that the bilinear form (α,ψ) will be important in the theory. Also, it suggests that the "adjoint" operator of A should have eigenfunctions of the form $e^{-\lambda\theta}$, $0 \leq \theta \leq r$, so that the "alternative" theorem could be stated in the following manner: the equation $(A-\lambda I)\varphi = \psi$ has a solution if and only if $(\alpha,\psi) = 0$ for all solutions α of the equation $(A^*-\lambda I)\alpha = 0$.

Proceeding from this formal point of view, as in Section 17, let us define $C^* = C([0,r],R^{n*})$, where R^{n*} is the n-dimensional vector space of row vectors, and for any α in C^*, φ in C, define

$$(\alpha,\varphi) = \alpha(0)\varphi(0) - \int_{-r}^{0}\int_{0}^{\theta} \alpha(\xi-\theta)[d\eta(\theta)]\varphi(\xi)d\xi \tag{21.1}$$

and try to determine an operator A^* with domain dense in C^* and range in C^* such that

$$(\alpha,A\varphi) = (A^*\alpha,\varphi), \text{ for } \varphi \text{ in } \mathscr{D}(A) \text{ and } \alpha \text{ in } \mathscr{D}(A^*). \tag{21.2}$$

If α has a continuous first derivative and φ is in $\mathscr{D}(A)$, then

$$
\begin{aligned}
(\alpha, A\varphi) &= \alpha(0)\int_{-r}^{0} [d\eta(\theta)]\varphi(\theta) - \int_{-r}^{0}\int_{0}^{\theta} \alpha(\xi-\theta)[d\eta(\theta)]\dot{\varphi}(\xi)d\xi \\
&= \alpha(0)\int_{-r}^{0} [d\eta(\theta)]\varphi(\theta) - \int_{-r}^{0} [\alpha(\xi-\theta)[d\eta(\theta)]\varphi(\xi)]_{0}^{\theta} \\
&\qquad + \int_{-r}^{0}\int_{0}^{\theta} \frac{d\alpha(\xi-\theta)}{d\xi}[d\eta(\theta)]\varphi(\xi)d\xi \\
&= [\int_{-r}^{0} \alpha(-\theta)d\eta(\theta)]\varphi(0) - \int_{-r}^{0}\int_{0}^{\theta} [-\frac{d\alpha(\xi-\theta)}{d\xi}][d\eta(\theta)]\varphi(\xi)d\xi \\
&= (A^*\alpha, \varphi)
\end{aligned}
$$

provided that we define

$$
A^*\alpha(s) = \begin{cases} -d\alpha(s)/ds, & 0 < s \leq r, \\ \int_{-r}^{0} \alpha(-\theta)d\eta(\theta), & s = 0. \end{cases} \tag{21.3}
$$

If the domain of A^* consists of all functions α in C^* which have a continuous first derivative such that

$$
-\frac{d\alpha(0)}{ds} = \int_{-r}^{0} \alpha(-\theta)d\eta(\theta)
$$

then A^* maps $\mathscr{D}(A^*)$ into C^* and $\mathscr{D}(A^*)$ is obviously dense in C^*. Hereafter, we will take (21.3) as the defining relationship for A^* so that (21.2) is satisfied and call A^* the "adjoint" of A relative to the bilinear form (21.1).

Even though it is not necessary for the following discussion, it is worthwhile for convenience in stating results to associate with (19.1) the adjoint differential equation of Section 17 defined by

$$
\dot{y}(\tau) = -\int_{-r}^{0} y(\tau-\theta)d\eta(\theta). \tag{21.4}
$$

Notice that this equation must in general be integrated in a backward direction and that existence and uniqueness of solutions holds.

If y is a solution of (21.4) on an interval $(-\infty,\sigma+r]$, then we let y^τ for each τ in $(-\infty,\sigma]$ designate the element in C^* defined by $y^\tau(\xi) = y(\tau+\xi)$, $0 \leq \xi \leq r$ (the "forward" restriction of y). Suppose that x is a solution of (19.1) on $\omega-r \leq t < \infty$ and y is a solution of (21.4) on $(-\infty,\sigma+r]$, $\sigma > \omega$. As was shown in Section 17

$$(y^t,x_t) = \text{constant} \tag{21.5}$$

for all t in $[\omega,\sigma]$. In particular, if y is defined on $(-\infty,\infty)$ then (21.5) holds for all $t \geq \omega$.

Another interesting property of (21.4) is the following: suppose $y(\alpha)$ is a solution of (21.4) on $(-\infty,r]$ with initial value α at zero and define $T^*(\tau)\alpha = y^\tau(\alpha)$, $-\infty < \tau \leq 0$. Then $T^*(\tau)$ has the same properties as the semigroup $T(t)$ associated with (19.1) and

$$\frac{dT^*(\tau)\alpha}{d\tau} = -A^*T^*(\tau)\alpha = -T^*(\tau)A^*\alpha \tag{21.6}$$

for all α in $\mathscr{D}(A^*)$.

These remarks should be sufficient motivation for the studying of the operator A^* and equation (21.4) in detail.

Lemma 21.1. λ is in $\sigma(A)$ if and only if λ is in $\sigma(A^*)$. The operator A^* has only point spectrum and for any λ in $\sigma(A^*)$, the generalized eigenspace of λ is finite dimensional.

Proof. The last part of the lemma is proved exactly as Lemma 20.1. A complex number λ is in $P\sigma(A^*)$ if and only if $\alpha(\theta) = e^{-\lambda\theta}b$ where b is a nonzero row vector satisfying

$$b[\lambda I - \int_{-r}^{0} e^{\lambda\theta} d\eta(\theta)] = b\Delta(\lambda) = 0;$$

that is, det $\Delta(\lambda) = 0$. This proves Lemma 21.1.

Lemma 21.2. A necessary and sufficient condition for the equation

$$(A-\lambda I)^k\varphi = \psi, \quad \psi \text{ in } C, \tag{21.7}$$

to have a solution φ in C or, equivalently, that ψ is in $\mathscr{R}(A-\lambda I)^k$, is that $(\alpha,\psi) = 0$ for all α in $\mathfrak{N}(A^*-\lambda I)^k$.

Proof. First, we introduce some notation. With the matrix $\Delta(\lambda)$ defined in (20.1), we define matrices P_j as

$$P_{j+1} = P_{j+1}(\lambda) = \frac{\Delta^{(j)}(\lambda)}{j!}, \quad \Delta^{(j)}(\lambda) = \frac{d^j\Delta(\lambda)}{d\lambda^j}, \; j = 0,1,2,\ldots, \tag{21.8}$$

and the matrices A_k of dimension $(kn) \times (kn)$ as

$$A_k = \begin{bmatrix} P_1 & P_2 & \cdots & P_k \\ 0 & P_1 & \cdots & P_{k-1} \\ \vdots & & & \\ 0 & 0 & \cdots & P_1 \end{bmatrix}. \tag{21.9}$$

If (21.7) is to have a solution, then necessarily

$$\left(\frac{d}{d\theta} - \lambda\right)^k\varphi(\theta) = \psi(\theta), \quad -r \le \theta \le 0$$

or

$$\varphi(\theta) = \sum_{j=0}^{k-1} \gamma_{j+1} \frac{\theta^j}{j!} e^{\lambda\theta} + \int_0^\theta e^{\lambda(\theta-\xi)} \frac{(\theta-\xi)^{k-1}}{(k-1)!}\psi(\xi)d\xi$$

where the γ_{j+1} are arbitrary n-dimensional column vectors which must be determined so that φ belongs to $\mathscr{D}(A-\lambda I)^k$. We now derive these conditions on the γ_j.

A simple induction argument on m shows that

$$\varphi^{(m)}(\theta) \overset{\text{def}}{=} \left(\frac{d}{d\theta} - \lambda\right)^m \varphi(\theta)$$

$$= \sum_{j=0}^{k-m-1} \gamma_{m+j+1} \frac{\theta^j}{j!} e^{\lambda\theta} + \int_0^\theta e^{\lambda(\theta-\xi)} \frac{(\theta-\xi)^{k-m-1}}{(k-m-1)!} \psi(\xi) d\xi$$

for $0 \leq m \leq k-1$.

Next, we observe that φ belongs to $\mathscr{D}(A-\lambda I)^k$ if and only if $\varphi^{(m)}$ belongs to $\mathscr{D}(A-\lambda I)$, $m = 0,1,2,\ldots,k-1$. Since a continuously differentiable function φ belongs to $\mathscr{D}(A)$ if and only if $\dot{\varphi}(0) = \int_{-r}^{0} [d\eta(\theta)]\varphi(\theta)$, it follows from the definition of the function $\varphi^{(m)}$ and the matrices P_{j+1} that $\varphi^{(m)}$, $m < k - 1$, belongs to $\mathscr{D}(A)$ if and only if

$$P_1\gamma_{m+1} + P_2\gamma_{m+2} + \ldots + P_{k-m}\gamma_k = \int_{-r}^{0} \int_0^\theta e^{\lambda(\theta-\xi)} \frac{(\theta-\xi)^{k-m-1}}{(k-m-1)!} [d\eta(\theta)]\psi(\xi)d\xi.$$

Since $\dot{\varphi}^{(k-1)}(0) = \lambda\gamma_k + \psi(0)$, it follows that $\varphi^{(k-1)}$ belongs to $\mathscr{D}(A)$ if and only if

$$P_1\gamma_k = -\psi(0) + \int_{-r}^{0} \int_0^\theta e^{\lambda(\theta-\xi)} [d\eta(\theta)]\psi(\xi)d\xi.$$

Introducing the notation

$$\gamma = \operatorname{col}(\gamma_1,\ldots,\gamma_k),$$

$$\alpha_j(s) = \frac{(-s)^{k-j}}{(k-j)!} e^{-\lambda s}, \quad 0 \leq s \leq r, \quad j = 1,2,\ldots,k,$$

$$\Psi = \operatorname{diag}(\alpha_1 I_n,\ldots,\alpha_k I_n),$$

where I_n is the $n \times n$ identity matrix, and using the above computations, we see that equation (21.7) has a solution if and only if γ satisfies the equation

$$A_k\gamma = -(\Psi,\psi).$$

The theory of matrix equations tells us that this equation has a solution if and only if $b(\Psi,\psi) = (b\Psi,\psi) = 0$ for all vectors b such that $bA_k = 0$. But, calculations very similar to the ones above show that α belongs to $\mathfrak{N}(A^*-\lambda I)^k$ if and only if $\alpha = b\Psi$ for some b satisfying $bA_k = 0$. This completes the proof of Lemma 21.2.

Since in the proof of Lemma 21.2, we have actually characterized $\mathfrak{N}(A-\lambda I)^k$ and $\mathfrak{N}(A^*-\lambda I)^k$ in a manner which is convenient for computations, we state this as

<u>Lemma 21.3.</u> $\mathfrak{N}(A-\lambda I)^k$ coincides with functions φ of the form

$$\varphi(\theta) = \sum_{j=0}^{k-1} \gamma_{j+1} \frac{\theta^j}{j!} e^{\lambda\theta}, \quad -r \leq \theta \leq 0,$$

where $\gamma = \text{col}(\gamma_1,\ldots,\gamma_k)$ satisfies $A_k\gamma = 0$ with A_k defined in (21.7), (21.8). Also, $\mathfrak{N}(A^*-\lambda I)^k$ coincides with functions ψ of the form

$$\psi(s) = \sum_{j=1}^{k} \beta_j \frac{(-s)^{k-j}}{(k-j)!} e^{-\lambda s}, \quad 0 \leq s \leq r$$

where $\beta = \text{row}(\beta_1,\ldots,\beta_k)$ satisfies $\beta A_k = 0$.

<u>Lemma 21.4.</u> For λ in $\sigma(A)$, let $\Psi_\lambda = \text{col}(\psi_1,\ldots,\psi_p)$, $\Phi_\lambda = (\varphi_1,\ldots,\varphi_p)$ be bases for $\mathfrak{M}_\lambda(A^*)$, $\mathfrak{M}_\lambda(A)$, respectively, and let $(\Psi_\lambda,\Phi_\lambda) = (\psi_i,\varphi_j)$, $j = 1,2,\ldots p$. Then $(\Psi_\lambda,\Phi_\lambda)$ is nonsingular and thus may be taken as the identity. The decomposition of C given by Lemma 20.1 may be written explicitly as

$$\begin{aligned}
&\varphi = \varphi^{P_\lambda} + \varphi^{Q_\lambda}, \quad \varphi^{P_\lambda} \text{ in } P_\lambda, \quad \varphi^{Q_\lambda} \text{ in } Q_\lambda, \\
&P_\lambda \overset{\text{def}}{=} \mathfrak{M}_\lambda(A) = \{\varphi \text{ in } C,\ \varphi = \Phi_\lambda b \text{ for some p-vector } b\}, \\
&Q_\lambda = \{\varphi \text{ in } C\colon (\Psi_\lambda,\varphi) = 0\}, \\
&\varphi^{P_\lambda} = \Phi_\lambda b, \quad b = (\Psi_\lambda,\varphi), \\
&\varphi^{Q_\lambda} = \varphi - \varphi^{P_\lambda}.
\end{aligned}$$

Proof. Lemma 21.2 implies that $Q_\lambda = \mathscr{R}(A-\lambda I)^k$ if k is the smallest integer for which $\mathfrak{M}_\lambda(A) = \mathfrak{N}(A-\lambda I)^k$. If there is a p-vector a such that $(\Psi,\Phi)a = (\Psi,\Phi a) = 0$, then Φa is in $\mathscr{R}(A-\lambda I)^k$ and $\mathfrak{N}(A-\lambda I)^k$ which from Lemma 20.1 implies $\Phi a = 0$ and, thus, $a = 0$. The remainder of the proof of the lemma is obvious.

It is also interesting to note that $(\Psi_\lambda,\Phi_\lambda) = I$ and $A^*\Psi_\lambda = B_\lambda^*\Psi_\lambda$, $A\Phi_\lambda = \Phi_\lambda B_\lambda$ implies $B_\lambda^* = B_\lambda$. In fact,

$$\begin{aligned}(\Psi_\lambda,A\Phi_\lambda) &= (\Psi_\lambda,\Phi_\lambda B_\lambda) = (\Psi_\lambda,\Phi_\lambda)B_\lambda = B_\lambda \\ &= (A^*\Psi_\lambda,\Phi_\lambda) = (B_\lambda^*\Psi_\lambda,\Phi_\lambda) = B_\lambda^*(\Psi_\lambda,\Phi_\lambda) = B_\lambda^*.\end{aligned}$$

Lemma 21.5. $\dim\mathfrak{M}_\lambda(A) =$ alg. multiplicity of λ.

For a proof of this lemma, see Levinger [J. Differential Equations, 1968].

Lemma 21.6. If $\lambda \neq \mu$, ψ is in $\mathfrak{M}_\mu(A^*)$ and φ is in $\mathfrak{M}_\lambda(A)$, then $(\psi,\varphi) = 0$.

The proof is left as an exercise.

We have already defined the generalized eigenspace of a characteristic value of (19.1) as the set $\mathfrak{M}_\lambda(A)$. If $\Lambda = \{\lambda_1,\dots,\lambda_p\}$ is a finite set of characteristic values of (19.1), we let $P = P_\Lambda$ be the linear extension of the $\mathfrak{M}_{\lambda_j}(A)$, $\lambda_j \in \Lambda$ and refer to this set as the generalized eigenspace of (19.1) associated with Λ. In a similar manner, we define $P^* = P_\Lambda^*$ as the generalized eigenspace of the adjoint equation (21.4) associated with Λ. If Φ, Ψ are bases for P_Λ, P_Λ^*, respectively, $(\Psi,\Phi) = I$, the identity, then

$$\begin{aligned}&C = P_\Lambda \oplus Q_\Lambda \\ &P_\Lambda = \{\varphi \text{ in } C\colon \varphi = \Phi b \text{ for some vector } b\} \\ &Q_\Lambda = \{\varphi \text{ in } C\colon (\Psi,\varphi) = 0\}\end{aligned} \tag{21.10}$$

and, therefore, for any φ in C,

$$\varphi = \varphi^{P_\Lambda} + \varphi^{Q_\Lambda}$$

(21.11)

$$\varphi^{P_\Lambda} = \Phi(\Psi,\varphi).$$

When this particular decomposition of C is used, we shall briefly express this by saying the C is decomposed by Λ.

22. ESTIMATES ON THE COMPLEMENTARY SUBSPACE

If C is decomposed by Λ, we know from Theorem 20.1 there is a constant matrix $B = B_\Lambda$ whose eigenvalues coincide with Λ such that

$$T(t)\varphi^{P_\Lambda} = \Phi e^{Bt}a, \quad \text{where } \varphi^{P_\Lambda} = \Phi a.$$

For the application of the theory of linear systems, we need to have an estimate for the solutions on the complementary subspace Q_Λ. Such an estimate requires a detailed knowledge of the spectrum of $T(t)$. In particular, we need to know the spectral radius of the semigroup $T(t)$ restricted to Q_Λ.

A first step in this direction is answered by the following result:

Lemma 22.1. If $T(t)$ is a strongly continuous semigroup on $[0,\infty)$ with infinitesimal generator A, then $P\sigma(T(t)) = e^{tP\sigma(A)}$ plus possibly $\{0\}$. More specifically, if $\mu = \mu(t) \neq 0$ is in $P\sigma(T(t))$ for some fixed t, then there is a point λ in $P\sigma(A)$ such that $e^{\lambda t} = \mu$. Furthermore, if $\{\lambda_n\}$ consists of all distinct points in $P\sigma(A)$ such that $e^{\lambda_n t} = \mu$, then $\mathfrak{N}(\mu I - T(t))^k$ is the closed linear extension of the linearly independent manifolds $\mathfrak{N}(\lambda_n I - A)^k$.

Proof. Lemma 22.1 is a special case of Theorem 16.7.2, p. 467 of Hille and Phillips for $k = 1$.

Exercise. Complete the proof for arbitrary k.

The spectral radius ρ of an operator T mapping a Banach space into itself is the smallest disk centered at the origin of the complex plane which contains $\sigma(T)$.

We also need the following

Lemma 22.2. If $T(t)$, $t \geq 0$ is a strongly continuous semigroup of operators of a Banach space $\mathscr{B}$ into itself, if for some $r > 0$, the spectral radius $\rho = \rho_{T(r)}$ is finite and $\neq 0$ and $\beta r = \log \rho$, then, for any $\gamma > 0$, there is a constant $K(\gamma) \geq 1$ such that

$$\|T(t)\varphi\| \leq K(\gamma)e^{(\beta+\gamma)t}\|\varphi\|, \quad \text{for all} \quad t \geq 0, \quad \varphi \quad \text{in} \quad \mathscr{B}.$$

Proof. Since $T(t)$ is strongly continuous, it is certainly bounded for each t and, in particular, $T(r)$ is bounded. It then follows from Reisz-Nagy, p. 425, that

$$\rho = e^{\beta r} = \lim_{n \to \infty} \|T^n(r)\|^{1/n}.$$

Therefore, for any $\gamma > 0$,

$$e^{-\gamma r} = \lim_{n \to \infty} e^{-(\beta+\gamma)r}\|T^n(r)\|^{1/n}$$

and there is a number N such that

$$e^{-(\beta+\gamma)nr}\|T^n(r)\| = (e^{-\gamma r} + \varepsilon_n)^n$$

where $e^{-\gamma r} + \varepsilon_n \leq L < 1$ for all $n \geq N$. Therefore,

$$e^{-(\beta+\gamma)nr}\|T^n(r)\| \to 0 \quad \text{as} \quad n \to \infty.$$

Since $T(t)$ is strongly continuous, there is a constant B such that $\|T(t)\| \leq B$ for $0 \leq t \leq r$. Define $K(\gamma)$ for any $\gamma > 0$ to be

$$K(\gamma) = Be^{|\beta+\gamma| r} \max_{n \geq 0} e^{-(\beta+\gamma)nr}\|T^n(r)\|.$$

If $0 \leq t \leq r$, then, for any φ in $\mathscr{B}$,

$$\|T(t)\varphi\| \leq \|T(t)\| \cdot \|\varphi\| \leq B\|\varphi\| \leq K(\gamma)e^{(\beta+\gamma)t}\|\varphi\|.$$

If $t \geq r$, then there is an integer n such that $nr \leq t < (n+1)r$ and for all φ

in $\mathscr{B}$,

$$\begin{aligned}\|T(t)\varphi\| &= \|T(t-nr)T(nr)\varphi\| \leq B\|T^n(r)\|\cdot\|\varphi\| \\ &= [Be^{-(\beta+\gamma)(t-nr)}e^{-(\beta+\gamma)nr}\|T^n(r)\|]e^{(\beta+\gamma)t}\|\varphi\| \\ &\leq K(\gamma)e^{(\beta+\gamma)t}\|\varphi\|.\end{aligned}$$

This completes the proof of the lemma.

If we now return to our original problem posed before the statement of Lemma 22.1, we obtain the following information.. Since $T(t)$ is compact for $t \geq r$, it follows that any μ in $\sigma(T(r))$, $\mu \neq 0$ is an element of $P\sigma(T(r))$ and that the only possible accumulation point in $\sigma(T(r))$ is zero. Furthermore, if $\mu \neq 0$ is in $P\sigma(T(r))$, then $\mathfrak{N}(\mu I - T(r))^k$ is of finite dimension for every k and $\mathfrak{M}_\mu(T(r))$ is finite dimension. These are well known properties of compact operators which can be found in Hille and Phillips, pp. 180-182. Lemmas 22.1 and 20.1 imply there are only a finite number of λ in $\sigma(A)$ such that $\operatorname{Re}\lambda \geq \beta$ for any given real number β. Consequently, if $\Lambda = \Lambda(\beta) = \{\lambda \text{ in } \sigma(A): \operatorname{Re}\lambda \geq \beta\}$ and C is decomposed by Λ, then there are constants $K > 0$, $\gamma > 0$, such that

$$\|T(t)\varphi^{Q_\Lambda}\| \leq Ke^{(\beta-\gamma)t}\|\varphi^{Q_\Lambda}\|, \quad t \geq 0.$$

We summarize these results in

Theorem 22.1. For any real number β, let $\Lambda = \Lambda(\beta) = \{\lambda \text{ in } \sigma(A): \operatorname{Re}\lambda \geq \beta\}$ and suppose C is decomposed by Λ as $C = P_\Lambda \oplus Q_\Lambda$. Then there exist positive constants K, γ such that

$$\|T(t)\varphi^{P_\Lambda}\| \leq Ke^{(\beta-\gamma)t}\|\varphi^{P_\Lambda}\|, \quad t \leq 0,$$

(22.1)

$$\|T(t)\varphi^{Q_\Lambda}\| \leq Ke^{(\beta-\gamma)t}\|\varphi^{Q_\Lambda}\|, \quad t \geq 0.$$

The first relation in (22.1) follows from Theorem 20.1 since we know that $T(t)$ can be defined on P_Λ for $-\infty < t < \infty$ and the eigenvalues of the corresponding matrix B_Λ associated with P_Λ coincides with the set Λ.

An important corollary of Theorem 22.1 concerning exponential asymptotic stability is

Corollary 22.1. If all of the roots of the characteristic equation (20.1) of (19.1) have negative real parts, then there exist positive constants K, γ such that

$$\|T(t)\varphi\| \leq Ke^{-\gamma t}\|\varphi\|, \quad t \geq 0,$$

for all φ in C.

The proof is obvious since by choosing $\beta = 0$ in Theorem 22.1 the set Λ is empty and P_Λ consists only of the zero element.

Theorems 20.1 and 22.1 give a very clear geometric picture of the behavior of the solutions of an autonomous functional differential equation. Of course, there are other interesting questions for linear systems which involve the concept of a fundamental system of solutions and, even more specifically, the closure of the set of functions $\{\Phi_\lambda, \lambda \text{ in } \sigma(A)\}$, where Φ_λ is a basis for $\mathfrak{M}_\lambda(A)$. Bellman and Cooke [1] and Pitt [1] have a thorough discussion of this question and among other things have proved under certain conditions that the solution of (19.1) for $t > 0$ can be represented as an infinite series involving these functions. They have also discussed the representation of an arbitrary Lipschitz continuous function as an infinite series of these functions. This aspect of functional differential equations is not discussed in these lectures.

23. AN EXAMPLE

Consider the scalar equation

$$\dot{x}(t) = -\frac{\pi}{2} x(t-1) \overset{\text{def}}{=} \int_{-1}^{0} [d\eta(\theta)]x(t+\theta) \tag{23.1}$$

where

$$\eta(\theta) = \begin{cases} 0, & \theta = -1 \\ -\frac{\pi}{2}, & -1 < \theta \leq 0 \end{cases}$$

and the adjoint system

$$\dot{y}(s) = \frac{\pi}{2} y(s+1). \tag{23.2}$$

The bilinear form is

$$(\psi,\varphi) = \psi(0)\varphi(0) - \frac{\pi}{2} \int_{-1}^{0} \psi(\xi+1)\varphi(\xi)d\xi \tag{23.3}$$

and the operators A, A^* are given by

$$[A\varphi](\theta) = \begin{cases} \dot{\varphi}(\theta) & , \quad -1 \leq \theta < 0 \\ -\frac{\pi}{2}\varphi(-1) , & \qquad \theta = 0 \end{cases}$$

$$[A^*\psi](s) = \begin{cases} -\dot{\psi}(s) & , \quad 0 < s \leq 1 \\ -\frac{\pi}{2}\psi(1) & , \qquad s = 0 \end{cases}$$

Moreover, φ is in $\mathfrak{N}(A-\lambda I)$ if and only if $\varphi(\theta) = e^{\lambda\theta}b$, $-r \leq \theta \leq 0$, where b is a constant and λ satisfies the characteristic equation

$$\lambda + \frac{\pi}{2} e^{-\lambda} = 0. \tag{23.4}$$

Also, ψ belongs to $\mathfrak{N}(A^*-\lambda I)$ if and only if $\psi(s) = e^{-\lambda s}c$, $0 \leq s \leq 1$, where c is a constant and λ satisfies (23.4).

It is easy to prove (and is left as an exercise) that equation (23.4) has two simple roots $\pm i\pi/2$ and the remaining roots have negative real parts.

Let $\Lambda = \{i\pi/2, -i\pi/2\}$ and then it is immediately obvious that

$$\Phi = (\varphi_1,\varphi_2), \quad \varphi_1(\theta) = \sin\frac{\pi}{2}\theta, \quad \varphi_2(\theta) = \cos\frac{\pi}{2}\theta, \quad -1 \leq \theta \leq 0, \tag{23.5}$$

is a basis for the generalized eigenspace $P = P_\Lambda$ of (23.1) associated with Λ and that

$$\Psi^* = \mathrm{col}(\psi_1^*,\psi_2^*), \quad \psi_1^*(s) = \sin\frac{\pi}{2}s, \quad \psi_2^*(s) = \cos\frac{\pi}{2}s, \quad 0 \leq s \leq 1,$$

is a basis for the generalized eigenspace P_Λ^* of (23.2) associated with Λ. We wish to decompose C by Λ. Also, we have seen that the transformations are simpler if $(\Psi^*,\Phi) = (\psi_j^*,\varphi_k)$, $j,k = 1,2$, is the identity matrix, and (ψ,φ) is defined in (23.3). If we compute this matrix, we see that it is not the identity. Therefore, we define a new basis Ψ for P_Λ^* by $\Psi = (\Psi^*,\Phi)^{-1}\Psi^*$ and then $(\Psi,\Phi) = I$. The explicit expression for the basis Ψ is

$$\begin{aligned} \Psi &= \mathrm{col}(\psi_1,\psi_2), \\ \psi_1(s) &= 2\mu\left[\sin\frac{\pi}{2}s + \frac{\pi}{2}\cos\frac{\pi}{2}s\right], \\ \psi_2(s) &= 2\mu\left[-\frac{\pi}{2}\sin\frac{\pi}{2}s + \cos\frac{\pi}{2}s\right], \quad \mu = 1/[1+\pi^2/4]. \end{aligned} \tag{23.6}$$

If we now decompose C by Λ and let $Q = Q_\Lambda$ for simplicity in notation, then any φ in C can be written as

$$(23.7)\qquad \begin{aligned} \varphi &= \varphi^P + \varphi^Q \\ \varphi^P &= \Phi b, \quad b = \mathrm{col}(b_1, b_2) \overset{\mathrm{def}}{=} (\Psi, \varphi) \\ b_1 &= \mu\pi\varphi(0) - \mu\pi \int_{-1}^{0} [\cos\tfrac{\pi}{2}\xi - \tfrac{\pi}{2}\sin\tfrac{\pi}{2}\xi]\varphi(\xi)d\xi \\ b_2 &= 2\mu\varphi(0) + \mu\pi \int_{-1}^{0} [\tfrac{\pi}{2}\cos\tfrac{\pi}{2}\xi + \sin\tfrac{\pi}{2}\xi]\varphi(\xi)d\xi. \end{aligned}$$

The explicit expression for b_1 and b_2 are obtained by simply substituting the expression for Ψ in (23.6) into (23.3).

From Theorem 22.1, we know that there are positive constants K, γ such that

$$(23.8)\qquad \|T(t)\varphi^Q\| \le Ke^{-\gamma t}\|\varphi^Q\|, \quad t \ge 0.$$

Consequently, the subspace P of C is asymptotically stable. More specifically, with A, Φ defined as above, we have

$$(23.9)\qquad A\Phi = \Phi B, \quad B = \begin{bmatrix} 0 & -\frac{\pi}{2} \\ \frac{\pi}{2} & 0 \end{bmatrix}$$

and, therefore, $T(t)\Phi = \Phi e^{Bt}$. Since $\varphi^Q = \varphi - \varphi^P$, $\varphi^P = \Phi b$, $b = (\Psi,\varphi)$, it follows from the estimate (23.8) that

$$\|T(t)\varphi - \Phi e^{Bt}b\| \to 0$$

exponentially as $t \to \infty$ for every φ in C, where $b = (\Psi,\varphi)$ is given explicitly in (23.7). That is, any solution of (23.1) approaches a periodic function of t given by $b_1 \sin \pi t/2 + b_2 \cos \pi t/2$ where b_1, b_2 satisfy (23.7).

In the (x,t)-space, it is very difficult to visualize this picture, but in C, everything is very clear. In the subspace P, $T(t)\varphi = \Phi e^{Bt}b$, the elements φ_1, φ_2 of Φ serve as a coordinate system in P and for any initial value Φb in P, we have $T(t+4)\Phi b = T(t)\Phi b$ since $\exp[B(t+4)] = \exp Bt$ and, in particular $T(4)\Phi b = \Phi b$; that is, the trajectories in C on P are closed curves. We can, therefore, symbolically represent the trajectories in C as in the accompanying figure. The pictorial representation of P by a (φ_1,φ_2)-plane is precise, but it should always be kept in mind that Q is an infinite dimensional space.

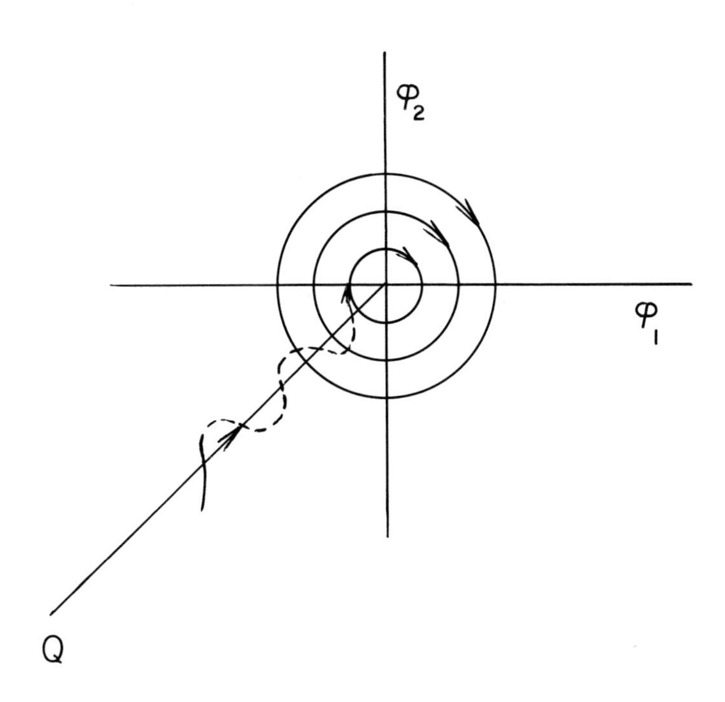

24. THE DECOMPOSITION IN THE VARIATION OF CONSTANTS FORMULA

Consider the equation

$$\dot{x}(t) = L(x_t) \tag{24.1}$$

and the nonhomogeneous system

$$\dot{x}(t) = L(x_t) + f(t) \tag{24.2}$$

where $f \in \mathcal{L}^1_{loc}([0,\infty), R^n)$. From our general results on linear systems, the general solution of (24.2) is

$$x_t = T(t-\sigma)\varphi + \int_\sigma^t T(t-s)X_0 f(s)ds \tag{24.3}$$

where

$$X_0(\theta) = \begin{cases} 0 & -r \leq \theta < 0 \\ I & \theta = 0 \end{cases}. \tag{24.4}$$

We now wish to obtain a decomposition of this integral equation according to the results on the decomposition of the homogeneous equation (24.1).

Suppose that $\Lambda = \{\lambda_1,\ldots,\lambda_p\}$ where each λ_j belongs to $\sigma(A)$ and that C is decomposed by Λ as $C = P \oplus Q$, Φ is a basis for the generalized eigenspace P of Λ, Ψ is a basis for the generalized eigenspace of the adjoint equation associated with Λ, $(\Psi,\Phi) = I$, and the matrix B is defined by the relations $A\Phi = \Phi B$, $A^*\Psi = B\Psi$. If the decomposition of any element φ of C is written as $\varphi = \varphi^P + \varphi^Q$, φ^P in P, φ^Q in Q, then $\varphi^P = \Phi(\Psi,\varphi)$. Suppose that x is the solution of (24.2) with initial value φ at σ, $x_t = x_t^P + x_t^Q$, and let us compute x_t^P directly from the above definition. To do this, we observe that if $y(t)$ is a solution of the adjoint equation on $(-\Phi,\Phi)$ and $x(t)$ is a solution of (24.2) for $t \geq 0$, then

$$(24.5) \qquad (y^t, x_t) = \int_\sigma^t y(s)f(s)ds + (y^\sigma, x_\sigma)$$

for all $t \geq 0$. Each row of the matrix $e^{-Bt}\Psi$, $\Psi(\theta) = e^{-B\theta}\Psi(0)$, $0 \leq \theta \leq r$, is a solution of the adjoint equation on $(-\infty, \infty)$ and, therefore,

$$(e^{-Bt}\Psi, x_t) = \int_\sigma^t e^{-Bs}\Psi(0)f(s)ds + (e^{-B\sigma}\Psi, \varphi)$$

and

$$\begin{aligned} x_t^P &\overset{\text{def}}{=} \Phi(\Psi, x_t) \\ &= \int_\sigma^t \Phi e^{B(t-s)}\Psi(0)f(s)ds + \Phi e^{B(t-\sigma)}(\Psi, \varphi) \\ &= \int_\sigma^t T(t-s)\Phi\,\Psi(0)f(s)ds + T(t-\sigma)\Phi(\Psi, \varphi) \\ &\overset{\text{def}}{=} \int_\sigma^t T(t-s)X_0^P\, f(s)ds + T(t-\sigma)\varphi^P \end{aligned}$$

where $X_0^P = \Phi\Psi(0)$.

This formula also shows another interesting property: namely,

$$(\Psi, x_t) = \int_\sigma^t e^{B(t-s)}\Psi(0)\; f(s)ds + e^{B(t-\sigma)}(\Psi, \varphi).$$

Therefore, if we let $y(t) = (\Psi, x_t)$ then

$$y(t) = e^{B(t-\sigma)}y(\sigma) + \int_\sigma^t e^{B(t-s)}\Psi(0)f(s)ds.$$

With x_t^P given as above, define $X_0^Q = X_0 - X_0^P = X_0 - \Phi\Psi(0)$. Then

$$x_t^Q = x_t - x_t^P = T(t-\sigma)\varphi^Q + \int_\sigma^t T(t-s)X_0^Q f(s)ds.$$

These results are summarized in

Theorem 24.1. Suppoce C is decomposed by Λ as $P \oplus Q$, Φ is a basis for P and Ψ is a basis for P^*. Then the solution $x(\sigma,\varphi)$ of (24.2) satisfies

$$x_t^P = T(t-\sigma)\varphi^P + \int_\sigma^t T(t-s)X_0^P f(s)ds \tag{24.6}$$

$$x_t^Q = T(t-\sigma)\varphi^Q + \int_\sigma^t T(t-s)X_0^Q f(s)ds, \quad t \geq \sigma,$$

where $X_0^P = \Phi\Psi(0)$, $X_0^Q = X_0 - \Phi\Psi(0)$. Furthermore, if $x_t^P = \Phi y(t)$ then

$$\dot{y}(t) = By(t) + \Psi(0)f(t). \tag{24.7}$$

Furthermore, if $\Lambda = \{\lambda \in \sigma(A)\colon \operatorname{Re}\lambda > \gamma\}$, then there are $\beta > 0$, $M > 0$ such that

$$\text{(24.8)} \quad \begin{aligned} &\text{(a)} \quad |T(t)\varphi^Q| \leq Me^{(\gamma-\beta)t}|\varphi^Q|, \quad \varphi^Q \in Q \\ &\text{(b)} \quad |T(t)X_0^Q| \leq Me^{(\gamma-\beta)t} \quad . \end{aligned}$$

Proof. Everything except (24.8b) has been proved. But this is obvious since $|T(t)X_0^Q| \leq M_1$ for $0 \leq t \leq r$ and for $t \geq r$, $T(t)X_0^Q$ belongs to Q.

As an example, consider

$$\dot{x}(t) = -\frac{\pi}{2}x(t-1) + f \tag{24.9}$$

and choose $\Lambda = \{+i\pi/2, -i\pi/2\}$. If we let $P = P(\Lambda)$, $Q = Q(\Lambda)$, $\varphi^P = \Phi(\Psi,\varphi)$ for any $\varphi \in C$, where Φ is defined in (23.5), Ψ in (23.6), $\varphi^Q = \varphi - \varphi^P$, $x_t = x_t^P + x_t^Q = \Phi y(t) + x_t^Q$, then

$$\dot{y}(t) = By(t) + \Psi(0)f$$

$$(24.10) \qquad x_t^Q = T(t-\sigma)\varphi^Q + \int_\sigma^t T(t-s)X_0^Q f(s)ds$$

where

$$\Psi(0) = (\pi\mu, 2\mu),$$

$$(24.11) \qquad \mu = [1 + \frac{\pi^2}{4}]^{-1} \qquad B = \begin{bmatrix} 0 & -\frac{\pi}{2} \\ \frac{\pi}{2} & 0 \end{bmatrix}.$$

If we let the components of y be y_1, y_2, then the first equation in (24.10) is given explicitly as

$$\dot{y}_1 = -\frac{\pi}{2} y_2 + \pi\mu f$$

$$\dot{y}_2 = \frac{\pi}{2} y_1 + 2\mu f.$$

If we let

$$z_1 = \frac{\pi}{2} y_1 + y_2, \quad y_1 = \mu(\frac{\pi}{2} z_1 + z_2)$$

$$(24.12) \qquad z_2 = y_1 - \frac{\pi}{2} y_2, \quad y_2 = \mu(z_1 - \frac{\pi}{2} z_2)$$

then

$$\dot{z}_1 = -\frac{\pi}{2} z_2$$

$$(24.13) \qquad \dot{z}_2 = \frac{\pi}{2} z_1 + 2f$$

and

$$\ddot{z}_1 + (\tfrac{\pi}{2})^2 z_1 = -\pi f. \tag{24.14}$$

It is interesting to see this second order equation (24.13) for a special type of forcing term, say, $f = f(x(t),x(t-1))$. From the definition of Φ and relations (24.12), (24.13), we have

$$x_t(0) - x_t^Q(0) = \Phi(0)y(t) = y_2(t) = \mu(z_1(t) + \dot{z}_1(t))$$
$$x_t(-1) - x_t^Q(-1) = \Phi(-1)y(t) = -y_1(t) = -\tfrac{2}{\pi}\mu(z_1(t) - \dot{z}_1(t)).$$

Consequently, if we neglect the terms $x_t^Q(0)$, $x_t^Q(t-1)$, the second order equation (24.14) becomes

$$\ddot{z}_1 + (\tfrac{\pi}{2})^2 z_1 = -\pi f[\mu(z_1 + \dot{z}_1), -\tfrac{2}{\pi}\mu(z_1 - \dot{z}_1)]. \tag{24.15}$$

In the applications to nonlinear oscillations, this equation plays a fundamental role.

Consider the system

$$\dot{x}(t) = L(x_t) \overset{\text{def}}{=} \int_{-r}^{0} [d\eta(\theta)]x(t+\theta) \tag{25.1}$$

and the perturbed linear system

$$\dot{x}(t) = L(x_t) + f(t) \tag{25.2}$$

where f belongs to $\mathscr{B}$, the class of bounded continuous functions mapping $(-\infty,\infty)$ into R^n with the topology of uniform convergence. For any σ in $(-\infty,\infty)$, we know from our variation of constants formula that the solution x of (25.2) with initial value x_σ at σ must satisfy

$$x_t = T(t-\sigma)x_\sigma + \int_\sigma^t T(t-s)X_0 f(s)ds, \quad t \geq \sigma. \tag{25.3}$$

We shall be interested in solutions of (25.3) which are bounded on $(-\infty,\infty)$. Recall that x is a solution of (25.3) on $(-\infty,\infty)$ if x is defined and continuous on $(-\infty,\infty)$ and for every σ in $(-\infty,\infty)$, relation (25.3) is satisfied.

Suppose $\Lambda = \Lambda_0 \cup \Lambda_1$, $\Lambda_0 = \{\lambda$ in $\sigma(A)$: Re $\lambda = 0\}$, $\Lambda_1 = \{\lambda$ in $\sigma(A)$: Re $\lambda > 0\}$, P_0, P_1 are the generalized eigenspaces of (25.1) associated with Λ_0, Λ_1, respectively, and that C is decomposed by Λ as $P_0 \oplus P_1 \oplus Q$. If

$$x_t = x_t^{P_0} + x_t^{P_1} + x_t^{Q},$$

then equation (25.3) is equivalent to

$$\begin{aligned} &\text{(a)} \quad x_t^{P_0} = T(t-\sigma)x_\sigma^{P_0} + \int_\sigma^t T(t-s)X_0^{P_0} f(s)ds, \\ (25.4) \quad &\text{(b)} \quad x_t^{P_1} = T(t-\sigma)x_\sigma^{P_1} + \int_\sigma^t T(t-s)X_0^{P_1} f(s)ds, \end{aligned}$$

(c) $x_t^Q = T(t-\sigma)x_\sigma^Q + \int_\sigma^t T(t-s)X_0^Q f(s)ds, \quad t \geq \sigma.$

Lemma 25.1. Equations (25.4b) and (25.4c) have unique solutions, $x_t^{P_1}$, x_t^Q which are bounded for $-\infty < t < \infty$ and these functions are given by

$$(25.5) \qquad \begin{aligned} x_t^Q &= \int_{-\infty}^t T(t-s)X_0^Q f(s)ds \quad , \\ x_t^{P_1} &= \int_{\infty}^t T(t-s)X_0^{P_1} f(s)ds, \quad -\infty < t < \infty. \end{aligned}$$

Furthermore, x_t^Q, $x_t^{P_1}$ are continuous linear functions on $\mathscr{B}$ in the sense that there is a constant $L > 0$ such that

$$(25.6) \qquad |x_t^{P_1}|, \ |x_t^Q| \leq L \sup_{s \text{ in } (-\infty,\infty)} |f(s)|, \quad -\infty < t < \infty.$$

Also, if f is ω-periodic, then

$$x_t^{P_1} = x_{t+\omega}^{P_1}, \quad x_t^Q = x_{t+\omega}^Q$$

for all t. If f is almost periodic in t, then $x_t^{P_1}$, x_t^Q are almost periodic in t with module contained in the module of f.

Proof. If x is a solution of (25.3) on $(-\infty,\infty)$, then we know that (25.4b), (25.4c) must hold for all σ in $(-\infty,\infty)$. Let us consider first the function x_t^Q. Since $\Lambda = \{\lambda \text{ in } \sigma(A)\colon \operatorname{Re} \lambda \geq 0\}$, we know there are $K > 0$, $\alpha > 0$ such that

$$|T(t)\varphi^Q| \leq Ke^{-\alpha t}|\varphi^Q|, \quad t \geq 0,$$

for all φ^Q in Q. Consequently,

$$|T(t-\sigma)x_\sigma^Q| \leq Ke^{-\alpha(t-\sigma)}|x_\sigma^Q|, \quad t \geq \sigma$$

and all σ in $(-\infty,\infty)$. If $|x_\sigma^Q|$ is assumed bounded uniformly in σ, it follows that $T(t-\sigma)x_\sigma^Q \to 0$ as $\sigma \to -\infty$ and this shows that x_t^Q satisfies (25.5). We also know that we can choose K,α above so that $|T(t)K_0^Q| \leq Ke^{-\alpha t}$, $t \geq 0$. The estimate (25.6) then follows immediately with $L = K/\alpha$. This shows that a bounded x_t^Q actually exists and it is unique by definition.

Since

$$x_t^Q = \int_{-\infty}^{0} T(-u)X_0^Q f(t+u)du,$$

it follows immediately that $f(t+\omega) = f(t)$ implies $x_{t+\omega}^Q = x_t^Q$ for all t. Also, this same expression and the above estimate on $|T(t)K_0^Q|$ shows that for any sequence $\{f(t+\tau_k)\}$ of functions which converges uniformly on $(-\infty,\infty)$, the functions $x_{t+\tau_k}^Q$ also converge uniformly on $(-\infty,\infty)$. Therefore, $f(t)$ almost periodic implies x_t^Q almost periodic and the module of x_t^Q belongs to the module of $f(t)$. This completes the proof of the lemma for x_t^Q.

For $x_t^{P_1}$, we proceed in a slightly different manner. Let Φ be a basis for P_1, Ψ be a basis for the generalized eigenspace of the adjoint equation associated with the set Λ_1, $(\Psi,\Phi) = I$, and let the matrix B whose eigenvalues coincide with Λ_1 be determined by $A\Phi = \Phi B$. For any φ in C, $\varphi^{P_1} = \Phi(\Psi,\varphi)$ and $T(t)^{P_1} = \Phi e^{Bt}(\Psi,\varphi)$. It then follows immediately from (25.4b) that

$$0 = \Phi[(\Psi,x_t) - e^{B(t-\sigma)}(\Psi,x_\sigma) - \int_\sigma^t e^{B(t-s)}\Psi(0)f(s)ds]$$

$$= \Phi[(\Psi,x_t) - e^{B(t-\sigma)}(\Psi,x_\sigma + \int_\sigma^t e^{B(\sigma-s)}X_0 f(s)ds)]$$

and therefore,

$$(\Psi,x_t) = e^{B(t-\sigma)}(\Psi,x_\sigma + \int_\sigma^t e^{B(\sigma-s)}X_0 f(s)ds).$$

Since the eigenvalues of B have positive real parts, there are positive constants K,α such that $|e^{Bt}a| \geq Ke^{\alpha t}|a|$ for any vector a and all $t \geq 0$. If $x_t^{P_1}$ is assumed bounded, we know that (Ψ, x_t) is bounded since $(\Psi, x_t) = (\Psi, x_t^{P_1})$. Thus, if $t \to \infty$ in the last equation, it follows that

$$(\Psi, x_\sigma + \int_\sigma^t e^{B(\sigma-s)} X_0 f(s)ds) \to 0 \quad \text{as} \quad t \to \infty.$$

Multiplication by Φ then yields

$$x_\sigma^{P_1} + \int_\sigma^t T(\sigma-s)X_0^{P_1} f(s)ds \to 0 \quad \text{as} \quad t \to \infty.$$

Finally, this implies

$$x_\sigma^{P_1} = \int_\infty^\sigma T(\sigma-s)X_0^{P_1} f(s)ds$$

for all σ in $(-\infty,\infty)$. This verifies the relation in (25.5) and the remainder of the proof is supplied in a manner similar to the above.

Theorem 25.1. A necessary and sufficient condition that (25.2) has a solution in $\mathscr{B}$ for every f in $\mathscr{B}$ is that system (25.1) have no characteristic values on the imaginary axis. Furthermore, the solution $\mathscr{K}f$ in $\mathscr{B}$ is unique, linear in f and there is a constant L such that

$$|\mathscr{K}f| \leq L \sup_{s \in (-\infty,\infty)} |f(s)|.$$

Furthermore, if f is ω-periodic then $\mathscr{K}f$ is ω-periodic and if f is almost periodic then $\mathscr{K}f$ is almost periodic with the module of $\mathscr{K}f$ equal to the module of f.

Proof. If (25.1) has no characteristic roots on the imaginary axis, then Lemma 25.1 yields all of the statements of the theorem except for the fact that the module of $\mathscr{K}f$ is the same as the module of f. Since $\mathscr{K}f$ satisfies (25.2) and is almost periodic, it follows that $d(\mathscr{K}f)/dt$ is almost periodic and obviously $d\mathscr{K}f/dt$ has the same module as $\mathscr{K}f$. Also, if the sequences $\{\mathscr{K}f(t+\sigma_k)\}$, $\{d\mathscr{K}f(t+\sigma_k)/dt\}$ converge uniformly on $(-\infty,\infty)$, then the sequence $\{f(t+\sigma_k)\}$ converges uniformly on $(-\infty,\infty)$ and thus the module of f belongs to the module of $\mathscr{K}f$. This completes the proof of sufficiency.

Now suppose that (25.2) has a solution in $\mathscr{B}$ for every f in $\mathscr{B}$ and suppose some elements of $\sigma(A)$ are on the imaginary axis. We show that this leads to a contradiction. Let C be decomposed by Λ as stated before Lemma 25.1. Let Φ be a basis for P_0, Ψ be a basis for the generalized eigenspace of the adjoint equation associated with Λ_0, $(\Psi,\Phi) = I$, and let $A\Phi = \Phi B$, where the eigenvalues of B coincide with Λ_0 and thus have real parts equal to zero. We shall show that there is an f in $\mathscr{B}$ such that every solution of (25.4a) is unbounded in $(-\infty,\infty)$. If $x_t^{P_0} = \Phi y(t)$, this is equivalent to showing the same thing for the equation

$$\dot{y}(t) = By(t) + \Psi(0)f(t). \tag{25.7}$$

It is clear that it is sufficient to consider the case where B is in Jordan canonical form and, thus, to consider Λ_0 to have only one element, which may be taken equal to zero. Consequently, we suppose $\Lambda_0 = \{0\}$. Suppose a is a nonzero row vector such that $aB = 0$. Then $a\Psi(0) \neq 0$; for otherwise, $ae^{-B(t-s)}\Psi(0) = ae^{-Bt}\Psi(s)$, $0 \leq s \leq r$, would be a nonzero solution of the adjoint equation which has zero initial value. For any solution y of (25.7), it follows that

$$a\dot{y}(t) = a\Psi(0)f(t).$$

If we let $f(t) = [a\Psi(0)]'$, then $|ay(t)|$ is unbounded as $t \to \infty$. Consequently, every solution of (25.7) is unbounded for this particular f. This completes the proof of the theorem.

Theorem 25.2. If $f(t+\omega) = f(t)$, then a necessary and sufficient condition that system (25.2) has a periodic solution of period ω is that

(25.8) $$\int_0^\omega y(t)f(t)dt = 0$$

for all ω-periodic solutions of the adjoint equation

$$\dot{y}(s) = -\int_{-r}^{0} y(s-\theta)d\eta(\theta).$$

Proof. From Lemma 25.1, it is clear that we only need to consider equation (25.4a). Furthermore, if $x_t^{P_0} = \Phi z(t)$, where the symbols are the same as in the proof of Theorem 25.1, then

(25.9) $$\dot{z}(t) = Bz(t) + \Psi(0)f(t),$$

where the eigenvalues of B have real parts equal to zero. From the theory of linear systems of ordinary differential equations, equation (25.9) has an ω-periodic solution if and only if

$$\int_0^\omega u(t)\Psi(0)f(t)dt = 0$$

for all ω-periodic solutions of $\dot{u} = -uB$. This implies

$$\int_0^\omega u_0 e^{-Bt}\Psi(0)f(t)dt = 0$$

for all u_0 for which $u_0 e^{-Bt}$ is ω-periodic. But $u_0 e^{-Bt}\Psi(0)$ for this set of u_0 coincides with the ω-periodic solutions of the adjoint equation. This completes the proof of Theorem 25.2.

26. THE SADDLE POINT PROPERTY

In this section, we consider the linear system

$$\dot{u}(t) = L(u_t) \tag{26.1}$$

and the perturbed linear system

$$\dot{x}(t) = L(x_t) + f(x_t) \tag{26.2}$$

where

$$L(\varphi) = \int_{-r}^{0} [d\eta(\theta)]\varphi(\theta), \tag{26.3}$$

η an $n \times n$ matrix function of bounded variation, and f satisfies the relations

$$\begin{aligned} &f(0) = 0 \\ &|f(\varphi) - f(\psi)| \leq \mu(\sigma)|\varphi-\psi| \end{aligned} \tag{26.4}$$

for $|\varphi|$, $|\psi| < \sigma$ and some continuous nondecreasing function $\mu(\sigma)$ with $\mu(0) = 0$.

It will also be assumed that the roots of the characteristic equation

$$\det \Delta(\lambda) = 0, \quad \Delta(\lambda) = \lambda I - \int_{-r}^{0} e^{\lambda\theta} d\eta(\theta) \tag{26.5}$$

have nonzero real parts. This latter assumption implies that the space C can be decomposed as

$$C = U \oplus S \tag{26.6}$$

where U is finite dimensional and the semigroup $T(t)$ generated by (26.1) can be defined on U for all $t \in (-\infty,\infty)$ and satisfies the relation

$$(26.7) \qquad \begin{aligned} |T(t)\varphi| &\leq Ke^{\alpha t}|\varphi|, \quad t \leq 0, \quad \varphi \in U. \\ |T(t)\varphi| &\leq Ke^{-\alpha t}|\varphi|, \quad t \geq 0, \quad \varphi \in S. \end{aligned}$$

For any $\varphi \in C$, we write $\varphi = \varphi^U + \varphi^S$, $\varphi^U \in U$, $\varphi^S \in S$. The decomposition of C as $U \oplus S$ defines two projection operators $\pi_U: C \to U$, $\pi_U U = U$, $\pi_S: C \to S$, $\pi_S S = S$, $\pi_S = I - \pi_U$.

Relations (26.7) and the fact that L is linear implies that system (26.1) behaves as a saddle point. More specifically, the orbits in C behave as shown in the accompanying figure.

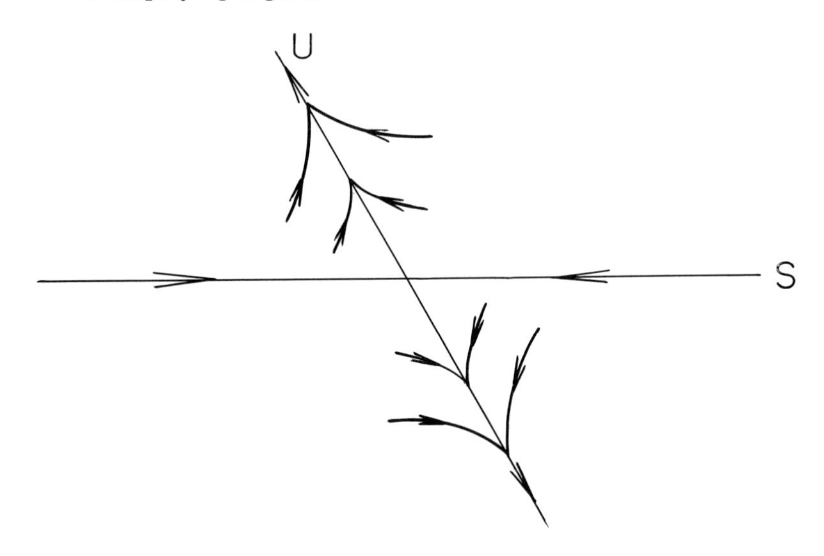

The set U is uniquely characterized as the set of initial values of those solutions of (26.1) which exist and remain bounded for $t \leq 0$. Relation (26.7) implies these solutions also approach zero exponentially as $t \to -\infty$. The set S is characterized as the set of initial values of those solutions of (26.1) which exist and remain bounded for $t \geq 0$. Any solution with initial value not on U or S is unbounded for $t \geq 0$ and if it exists for $t \leq 0$, it is also unbounded.

It is natural to ask if the solutions of (26.2) have the same qualitative behavior near $x = 0$ as the solutions of (26.1). Of course, the meaning of qualitative behavior must be defined very carefully. If one says that (26.2) has the same qualitative behavior as (26.1) near $x = 0$ if the orbits of (26.2) can be mapped homeomorphically into the orbits of (26.1) as in ordinary differential equations, then the following example suggests such a definition is too strong.

Consider the scalar system

$$\dot{y}(t) = -\beta y(t-1)[1+y(t)], \quad \beta > 0.$$

The constant function $y(t) = -1$, $t \in (-\infty,\infty)$, is an equilibrium point of this equation. If we let $x(t) = y(t) + 1$, then x satisfies the equation

$$\dot{x}(t) = \beta x(t) - \alpha x(t)x(t-1)$$

which is a special case of system (26.2) with $L(\varphi) = \alpha\varphi(0)$, $r = 1$, $f(\varphi) = -\alpha\varphi(0)\varphi(-1)$. The sets U, S are given by

$$U = \{\varphi\colon \varphi(\theta) = e^{\beta\theta}\varphi(0), \quad -1 \le \theta \le 0\}$$

$$S = \{\varphi\colon \varphi(\theta) = \psi(\theta) - e^{\beta\theta}\psi(0), \quad -1 \le \theta \le 0, \quad \psi \in C\}.$$

If $\varphi \in S$, then $T(t)\varphi = 0$ for $t \ge 1$. Therefore, the orbits in C have the form shown in the accompanying figure:

It is not unreasonable to suspect that the orbits for the nonlinear equation will never intersect. Consequently, it seems to be very difficult to obtain a precise relation between the orbits of the two systems.

On the other hand, we can show that some of the important properties of the trajectories are preserved. More specifically, we show below that the set of initial values of those solutions of (26.2) which exist and remain in a δ-neighborhood of $x = 0$ for $t \leq 0$ is homeomorphic to a neighborhood in U of zero and these solutions approach zero exponentially as $t \to -\infty$. The same result is proved for $t \geq 0$ and S. Any other solution must leave a neighborhood of zero with increasing t and if it exists for $t \leq -r$, it must also leave a neighborhood of zero.

Suppose K, α are defined in (26.7) and $x(\varphi)$ is the solution of (26.2) with initial value φ at zero. For any $\delta > 0$, let $B_\delta = \{\varphi \in C: |\varphi| \leq \delta\}$ and

$$(26.8)\qquad \begin{aligned} \mathscr{S}_\delta &= \{\varphi \in C: \varphi^S \in B_{\delta/2K},\ x_t(\varphi) \in B_\delta,\ t \geq 0\}, \\ \mathscr{U}_\delta &= \{\varphi \in C: \varphi^U \in B_{\delta/2K},\ x_t(\varphi) \in B_\delta,\ t \leq 0\}. \end{aligned}$$

If Γ is a subset of C which contains zero, we say Γ <u>is tangent to</u> S <u>at zero</u> if $|\pi_U\varphi|/|\pi_S\varphi| \to 0$ as $\varphi \to 0$ in Γ. Similarly, Γ <u>is tangent to</u> U <u>at zero</u> if $|\pi_S\varphi|/|\pi_U\varphi| \to 0$ as $\varphi \to 0$ in Γ.

We can now state the main result of this section.

<u>Theorem 26.1.</u> With the notation as above, there is a $\delta > 0$ such that π_S is a homeomorphism from the set $\mathscr{S}_\delta$ onto $S \cap B_{\delta/2K}$ and $\mathscr{S}_\delta$ is tangent to S at zero. Also, π_U is a homeomorphism from the set $\mathscr{U}_\delta$ onto $U \cap B_{\delta/2K}$ and $\mathscr{U}_\delta$ is tangent to U at zero. Furthermore, there are positive constants M, γ such that

$$(26.9)\qquad \begin{aligned} |x_t(\varphi)| &\leq Me^{-\gamma t}|\varphi|, \quad t \geq 0, \quad \varphi \text{ in } \mathscr{S}_\delta, \\ |x_t(\varphi)| &\leq Me^{\gamma t}|\varphi|, \quad t \leq 0, \ \varphi \text{ in } \mathscr{U}_\delta. \end{aligned}$$

Finally, if $f(\varphi)$ has a continuous Frechet derivative with respect to φ and $h_S: S \cap B_{\delta/2K} \to \mathcal{S}_\delta$, $h_U: U \cap B_{\delta/2K} \to \mathcal{U}_\delta$ are defined by $h_S\varphi = \pi_S^{-1}\varphi$, $\varphi \in S \cap B_{\delta/2K}$, $h_U\varphi = \pi_U^{-1}\varphi$, $\varphi \in U \cap B_{\delta/2K}$, then h_S and h_U have continuous Frechet derivatives.

Using the decomposition (26.6), the solution $x = x(\varphi)$ of (26.2) can be written as

$$
(26.10) \quad
\begin{aligned}
&(a) \quad x_t = x_t^S + x_t^U \\
&(b) \quad x_t^S = T(t-\sigma)x_\sigma^S + \int_\sigma^t T(t-s)X_0^S f(x_s)ds \\
&(c) \quad x_t^U = T(t-\sigma)x_\sigma^U + \int_\sigma^t T(t-s)X_0^U f(x_s)ds
\end{aligned}
$$

for any $\sigma \in (-\infty,\infty)$ and K, α can be chosen so that

$$
(26.11) \quad
\begin{aligned}
|T(t)X_0^U| &\leq Ke^{\alpha t}, \quad t \leq 0 \\
|T(t)X_0^S| &\leq Ke^{-\alpha t}, \quad t \geq 0.
\end{aligned}
$$

The reasoning from this point will use only the integral equations (26.10), the estimates (26.7), (26.11) and the fact that $T(t)\varphi^U$, $T(t)X_0^U$ are defined for all $t \in (-\infty,\infty)$. Therefore, the proof will have implications for certain types of equations of neutral type.

The following lemmas are needed in the proofs.

Lemma 26.1. With the above notation, for any solution x of (26.2) which exists and is bounded for $t \geq 0$, there is a φ^S in S such that

$$
(26.12) \quad
\begin{aligned}
x_t = T(t)\varphi^S &+ \int_0^t T(t-s)X_0^S f(x_s)ds \\
&+ \int_\infty^0 T(-s)X_0^U f(x_{t+s})ds
\end{aligned}
$$

for $t \geq 0$. For any solution x of (26.2) which exists and is bounded on $(-\infty,0]$, there is a φ^U in U such that

$$x_t = T(t)\varphi^U + \int_0^t T(t-s)X_0^U f(x_s)ds \tag{26.13}$$
$$+ \int_{-\infty}^0 T(-s)X_0^S f(x_{t+s})ds$$

for $t \leq 0$. Conversely, any solution of (26.12) bounded on $[0,\infty)$ and any solution of (26.13) bounded on $(-\infty,0]$ is a solution of (26.2).

Proof. Suppose x is a solution of (26.2) which exists for $t \geq 0$ and $|x_t| \leq M$ for $t \geq 0$. There is an $L \geq 0$ such that $|\pi_U\varphi| \leq L|\varphi|$ for all $\varphi \in C$ and, thus, $|\pi_U x_t| \leq LM$ for all $t \geq 0$. Since f maps bounded sets into bounded sets, there is a constant N such that $|f(x_t)| \leq N$, $t \geq 0$. For any σ in $[0,\infty)$, x_t^U satisfies (26.10c). Since (26.9) is satisfied and x_σ^U is bounded in σ, it follows that $T(t-\sigma)x_\sigma^U \to 0$ as $\sigma \to \infty$. Also, for $t \leq \sigma$,

$$\left|\int_\sigma^t T(t-s)X_0^U f(x_s)ds\right| \leq \frac{KLN}{\alpha}.$$

Therefore, the integral $\int_\infty^t T(t-s)X_0^U f(x_s)ds$ exists. Letting $\sigma \to \infty$ in (26.10c), it follows that

$$x_t^U = \int_\infty^t T(-s)X_0^U f(x_s)ds.$$

This proves relation (26.12). Relation (26.13) is proved in a similar manner. The last statement of the lemma is verified by direct computation.

Lemma 26.2. Suppose α, γ, K, L, M are nonnegative constants and u is a nonnegative bounded solution of either the inequality

$$u(t) \leq Ke^{-\alpha t} + L\int_0^t e^{-\alpha(t-s)}u(s)ds + M\int_0^\infty e^{-\gamma s}u(t+s)ds, \quad t \geq 0$$

or

$$u(t) \leq Ke^{\alpha t} + L\int_t^0 e^{\alpha(t-s)}u(s)ds + M\int_{-\infty}^0 e^{\gamma s}u(t+s)ds, \quad t \leq 0.$$

If

$$\beta \overset{\text{def}}{=} \frac{L}{\alpha} + \frac{M}{\gamma} < 1,$$

then

$$u(t) \leq (1-\beta)^{-1}Ke^{-[\alpha-(1-\beta)^{-1}L]|t|}.$$

The proof of this lemma is left as an exercise.

Proof of Theorem 26.1. From Lemma 26.1, for any solution x of (26.2) which is bounded on $[0,\infty)$, there is a φ^S in S such that x satisfies (26.12). We first discuss the solution of (26.12) for any φ^S sufficiently small. Suppose K, α are the constants in (26.9), (26.11) and $\mu(\sigma)$, $\sigma \geq 0$, is the function given in (26.4). Choose $\delta > 0$ so small that

$$4K\mu(\delta) < \alpha, \quad 8K^2\mu(\delta) < 3\alpha. \tag{26.14}$$

With choice of δ, define $\mathcal{G}(\delta)$ as the set of continuous functions $y\colon [0,\infty) \to C$ such that $|y| \overset{\text{def}}{=} \sup_{0\leq t<\infty}|y_t| \leq \delta/2$, $y_0^S = 0$. The set $\mathcal{G}(\delta)$ is a closed bounded subset of the Banach space $C([0,\infty),C)$ of all bounded continuous functions mapping $[0,\infty)$ into C with the uniform topology. For any y in $\mathcal{G}(\delta)$ and any φ^S in S, $|\varphi^S| \leq \delta/2K$, define the transformation $\mathcal{P} = \mathcal{P}(\varphi^S)$ taking $\mathcal{G}(\delta)$ into $C([0,\infty),C)$ by

$$\begin{aligned}(\mathcal{P}y)_t &= \int_0^t T(t-s)X_0^S f(y_s + T(s)\varphi^S)ds \\ &\quad + \int_\infty^0 T(-s)X_0^U f(y_{t+s} + T(t+s)\varphi^S)ds\end{aligned} \tag{26.15}$$

for $t \geq 0$. It is easy to see that $\mathscr{P}y \in C([0,\infty),C)$ and $(\mathscr{P}y)_0^S = 0$. Also, $|y_t + T(t)\varphi^S| \leq \delta$ for all $t \geq 0$. Consequently,

$$|(\mathscr{P}y)_t| \leq \frac{2K}{\alpha}\mu(\delta)\delta < \delta/2$$

and $\mathscr{P}$: $\mathscr{G}(\delta) \to \mathscr{G}(\delta)$. Furthermore,

$$|(\mathscr{P}y)_t - (\mathscr{P}z)_t| \leq \frac{2K}{\alpha}\mu(\delta)|y-z| \leq \frac{1}{2}|y-z|$$

for $t \geq 0$, $y,z \in \mathscr{G}(\delta)$ and $\mathscr{P}$ is a uniform contraction on $\mathscr{G}(\delta)$. Thus, $\mathscr{P}$ has a unique fixed point $y^* = y^*(\varphi^S)$ in $\mathscr{G}(\delta)$. The function $x_t^* = y_t^* + T(t)\varphi^S$ obviously satisfies (26.12) and is the unique solution of (26.12) with $|y_t| \leq \delta/2$ and $x_0^S = \varphi^S$. The fact that $\mathscr{P}$ is a uniform contraction on $\mathscr{G}(\delta)$ implies that $y^*(\varphi^S)$ and therefore $x^*(\varphi^S)$ are continuous in φ^S.

With x^* defined as above, let $x^* = x^*(\varphi^S)$, $\tilde{x}^* = x^*(\tilde{\varphi}^S)$. Then

$$|x_t^* - \tilde{x}_t^*| \leq Ke^{-\alpha t}|\varphi^S - \tilde{\varphi}^S| + K\mu(\delta)\int_0^t e^{-\alpha(t-s)}|x_s^* - \tilde{x}_s^*|ds$$

$$+ K\mu(\delta)\int_0^\infty e^{-\alpha s}|x_{t+s}^* - \tilde{x}_{t+s}^*|ds$$

for $t \geq 0$. If Lemma 26.2 is applied to this relation, then

$$|x_t^*(\varphi^S) - x_t^*(\tilde{\varphi}^S)| \leq 2Ke^{-\alpha t/2}|\varphi^S - \tilde{\varphi}^S|, \quad t \geq 0.$$

Since $x^*(0) = 0$, this relation implies (26.9) is satisfied.

The above arguments has also shown that

$$\mathscr{S}_\delta = \{\varphi \in C\colon \varphi = x_0^*(\varphi^S),\ \varphi^S \text{ in } S,\ |\varphi^S| \leq \delta/2K\}.$$

If $h_S\colon S \cap B_{\delta/2K} \to \mathscr{S}_\delta$ is defined by $h_S\varphi^S = x_0^*(\varphi^S)$, then h_S is continuous and

$$h_S(\varphi^S) = \varphi^S + \int_\infty^0 T(-s)X_0^U f(x_s^*(\varphi^S))ds.$$

Also, $|h_S(\varphi^S) - h_S(\tilde{\varphi}^S)| \geq |\varphi^S - \tilde{\varphi}^S|/2$ for all φ^S, $\tilde{\varphi}^S$ in $S \cap B_{\delta/2K}$, and, thus, h_S is one-to-one. Since $h_S^{-1} = \pi_S$ is continuous, it follows that h_S is a homeomorphism.

From the fact that $x_0^*(0) = 0$, we also have

$$|\pi_U x_0^*(\varphi^S)| \leq \frac{4K^2}{\alpha} \mu(2K|\varphi^S|)|\varphi^S|$$

and this shows that $\mathcal{S}_\delta$ is tangent to S at zero.

If f has a continuous Frechet derivative $f'(\varphi)$ and satisfies (26.4), then $f'(0) = 0$. From (26.15), it follows that the derivative $\mathcal{P}'(y)$ of $\mathcal{P}y$ with respect to φ^S evaluated at ψ^S in S is

$$(\mathcal{P}'(y)\psi^S)_t = \int_0^t T(t-s)X_0^S f'(y_s + T(s)\varphi^S)T(s)\psi^S ds$$
$$+ \int_\infty^0 T(-s)X_0^U f'(y_{t+s} + T(t+s)\varphi^S)T(t+s)\psi^S ds, \quad t \geq 0.$$

Since $|T(s)\psi^S| \leq K|\psi^S|$ and $\mu(\delta)$ satisfies (26.14), it follows that

$$|(\mathcal{P}'(y)\psi^S)_t| \leq \frac{2K^2\mu(\delta)}{\alpha}|\psi^S| < \frac{3}{4}|\psi^S|, \quad t \geq 0.$$

Using the fact that the mapping $\mathcal{P}$ is a uniform contraction on $\mathcal{G}(\delta)$, one obtains the differentiability of $h_S(\varphi^S)$ with respect to φ^S. The argument for $\mathcal{U}_\delta$ is applied similarly to the above to complete the proof of Theorem 26.1.

<u>Corollary 26.1.</u> Under the hypothesis of Theorem 26.1, there is a $\delta > 0$ such that each solution of (26.2) with initial value in B_δ either approaches zero as $t \to \infty$ (and then exponentially) or leaves B_δ for some finite time. Any solution with initial value in B_δ which is defined for $t \leq -r$ must either approach zero as $t \to -\infty$ or leave B_δ for some finite negative time.

Proof. There is a $k \geq 1$ such that $|\varphi^S| \leq k|\varphi|$ for all φ in C. Suppose δ is given as in Theorem 26.1 and choose $0 < \delta_1 \leq \delta/2Kk$. This δ_1 serves for the δ of the corollary. A similar argument applies to the last statement of the corollary.

Theorem 26.2. Suppose the hypotheses of Theorem 26.1 are satisfied, $\Phi = (\varphi_1,\ldots,\varphi_d)$ is a basis for U and let $\pi_U\varphi = \Phi b$ where b is a d-vector uniquely defined by φ. If p is an arbitrary real number, $0 < p < 1$, then there exist $\delta_0 > 0$ and a positive definite quadratic form $V(\varphi)$ in the components of b such that, for any δ, $0 < \delta \leq \delta_0$, $\dot{V}(\varphi) > 0$ if $|\varphi^U| \geq p\delta$, $\varphi \in B_\delta$, where $\dot{V}(\varphi)$ is the upper right hand derivative of V along the solutions of (26.2).

Proof. If we let $X_0^U = \Phi C$, $T(t)\Phi = \Phi e^{Bt}$, $\varphi^U = \Phi b$, $x_t^U = \Phi y(t)$, then relation (26.10) implies

$$y(t) = e^{Bt}b + \int_0^t e^{B(t-s)}Cf(x_s)ds,$$

where B has all eigenvalues with positive real parts. Thus,

$$\dot{y}(t) = By(t) + Cf(x_t).$$

Define $V(\varphi) = b'Ab$, where b' is the transpose of b and the $d \times d$ positive definite matrix A is chosen so that

$$B'A + AB = I.$$

If $g(\varphi)$ is defined by $g(\varphi) = Cf(\varphi)$, then

$$\dot{V}(x_t) = y'y + 2g'Ay.$$

If $|x_t^U| = |\Phi y| \geq p\delta$, then $|y| \geq p_1\delta$ where $p_1 > 0$. Choose δ_0 so that $\eta(\delta_0) <$

$kp_1/4|A|$, where $k > 0$ satisfies $k|y|^2 \le y'y$. Thus, as long as $|x_t| \le \delta \le \delta_0$

$$\begin{aligned}\dot{V}(x_t) &\ge k|y|^2 - 2|A|\eta(\delta)|x_t|\cdot|y| \\ &\ge |y|^2\left(1 - \frac{2|A|\eta(\delta)|x_t|}{k|y|}\right) \\ &\ge |y|^2\left(1 - \frac{2|A|\eta(\delta)}{kp_1}\right) \\ &\ge \frac{1}{2}|y|^2.\end{aligned}$$

This proves Theorem 26.2.

This last theorem holds even if some eigenvalues of (26.1) are on the imaginary axis. It is easily checked that there is no change in the argument.

27. A FIXED POINT THEOREM FOR CONES

Let E be a Banach space. A set $M \subset E$ is convex if $\varphi, \psi \in M$ implies $\lambda\varphi + (1-\lambda)\psi \in M$, $0 \leq \lambda \leq 1$. A set $K \subset E$ is a cone if

(i) K is closed and convex,

(ii) if φ is in K, then $\lambda\varphi \in K$, $\lambda \geq 0$,

(iii) for any $\varphi \neq 0$ in E, both φ and $-\varphi$ cannot belong to K.

A truncated cone is the intersection of a cone with a convex neighborhood of zero. The neighborhood need not be closed.

If K is a cone and $\varphi, \psi \in E$, we say $\varphi \geq \psi$ if $\varphi - \psi \in K$.

Lemma 27.1. Suppose K is a cone.

(i) If $\gamma > 0$, $u \in E$, $\varphi \in K$ and $\varphi - \gamma u \notin K$, then $\varphi \geq tu$ implies $t < \gamma$.

(ii) If $u \in E$, $K+u = \{\psi \in E: \psi = \varphi + u, \varphi \in K\}$ and $-u \notin K$, then $\mathrm{dist}(K+u, 0) > 0$.

(iii) If $u \in E$, $-u \notin K$, then for any $\varphi \in E$, there is a $\gamma > 0$ such that $\varphi - \gamma u \notin K$.

Proof. (i) We may assume $t > 0$. If $\varphi - tu \in K$, then convexity implies $\lambda(\varphi - tu) + (1-\lambda)\varphi = \varphi - \lambda tu \in K$, $0 \leq \lambda \leq 1$. If $\gamma \leq t$, then for $\lambda = \gamma/t$, we have a contradiction.

(ii) This is immediate from the closure of K and the fact that $-u \notin K$.

(iii) If there is a sequence $t_n \geq 0$, $t_n \to \infty$ as $n \to \infty$ such that $\varphi - t_n u \in K$, $n = 1,2,\ldots$, then $(\varphi/t_n) - u \in K$, $n = 1,2,\ldots$. But this sequence converges to $-u$. Closure of K implies $-u \in K$, which is a contradiction.

If K is a cone (or a truncated cone) in E, any $\varphi \in K$ will be called a positive vector. If A is a mapping defined on E or a part of E, $A: K \to K$, then A will be called a positive operator. If $A: \mathscr{D}(A) \subset E \to E$, A linear or nonlinear, then $\varphi \neq 0$ is called an eigenvector of A if there is a μ such that

$A\varphi = \mu\varphi$. If A is a positive operator and $\varphi \in K$ is an eigenvector of A, then φ will be called a positive eigenvector. A mapping $A: \mathscr{D}(A) \to E$ is a compact mapping if it takes bounded sets in $\mathscr{D}(A)$ into relative compact sets of E. A mapping $A: \mathscr{D}(A) \to E$ is completely continuous if it is compact and continuous. In the following, we let

$$B_r = \{\varphi \in E: |\varphi| < r\}, \quad S_r = \{\varphi \in E, |\varphi| = r\}, \quad \overline{B}_r = B_r \cup S_r.$$

Lemma 27.2. If K is a cone, A is a positive completely continuous operator and, for some $r > 0$,

$$\inf_{\varphi \in K \cap S_r} |A\varphi| > 0, \tag{27.1}$$

then A has a positive eigenvector $\varphi \in K \cap S_r$.

Proof. Suppose $u \neq 0$ is some fixed element in K. For any $\varphi \in K \cap \overline{B}_r$, let

$$A_1\varphi = |\varphi| A(\frac{r}{|\varphi|}\varphi) + (r - |\varphi|)u.$$

It is clear that A_1 is a positive operator. For any fixed r_0, $0 \leq r_0 \leq r$, the

$$\inf_{\varphi \in K \cap S_{r_0}} |r_0 A(\frac{r}{r_0}\varphi) + (r-r_0)u|$$

is a continuous function of r_0. For $r_0 = r$, we have assumed this is positive. For $r_0 < r$, it is positive from Lemma 27.1. Therefore

$$\inf_{\varphi \in K \cap \overline{B}_r} |A_1\varphi| > 0.$$

If we define

$$B\varphi = \frac{A_1\varphi}{|A_1\varphi|}, \quad \varphi \in K \cap \overline{B}_r,$$

then $B: K \cap \bar{B}_r \to K \cap S_r \subset K \cap \bar{B}_r$. Obviously, B is completely continuous. Therefore, Schauder's theorem implies the existence of a fixed point of B in $K \cap S_r$, that is, a $\varphi \in K \cap S_r$ such that $B\varphi = \varphi$, or $A_1\varphi = |A_1\varphi| \cdot \varphi$ or $A\varphi = (|A_1\varphi|/r)\varphi$. This proves the lemma.

Example. Let $E = C([0,1],R^1)$ and define $A: E \to E$ by $A\varphi(t) = \int_0^t \varphi(s)ds$, $\varphi \in E$. The operator A is completely continuous. Let $K = \{\varphi \in E: \varphi(t) \geq 0, 0 \leq t \leq 1\}$. K is a cone and $A: K \to K$. On the other hand, K has no positive eigenvectors. In fact, if this were so, then there would be a positive $\lambda > 0$ and a $\varphi \in E$, $\varphi(0) = 0$, such that $\int_0^t \varphi(s)ds = \varphi(t)/\lambda$, $0 \leq t \leq 1$. Thus, $\varphi(t) = e^{\lambda t}\varphi(0) = 0$ for all t. Notice that condition (27.1) is not satisfied.

The next result is a generalization of Lemma 27.2.

Theorem 27.1. Suppose K is either a cone or a truncated cone and ∂G is the boundary of an open bounded set $G \subset E$ with $0 \in G$. If $A: \partial G \cap K \to K$ is completely continuous and

$$\inf_{\varphi \in \partial G \cap K} |A\varphi| > 0 \tag{27.2}$$

then A has an eigenvector on $\partial G \cap K$.

Proof. Suppose first that E is finite dimensional. Without loss of generality, we may assume K is solid, that is, K contains interior points. We simply restrict ourselves to the relative topology. Hypothesis (27.2) becomes $\min_{\varphi \in \partial G \cap K} |A\varphi| > 0$. Let T be the closed convex hull of $A(\partial G \cap K)$. If $\varphi, \psi \in K$ and there exists a λ, $0 \leq \lambda \leq 1$ such that $\lambda\varphi + (1-\lambda)\psi = 0$, then either φ or ψ is zero. Therefore, T cannot contain zero since it would imply there is a $\varphi \in \partial G \cap K$ such that $A\varphi = 0$. Suppose the dimension of T is k.

T belongs to a k-dimensional hyperplane E_1 in E. Introduce in E_1 a coordinate system $\xi_1, \ldots, \xi_k$. Then for any $y \in E$, there is a unique set of α_j such that $y = \sum \alpha_j \xi_j$. Consequently, for any x in $\partial G \cap K$, $Ax = \sum \alpha_j(x) y_j$ and the mapping A on $\partial G \cap K$ is specified by giving the continuous functions

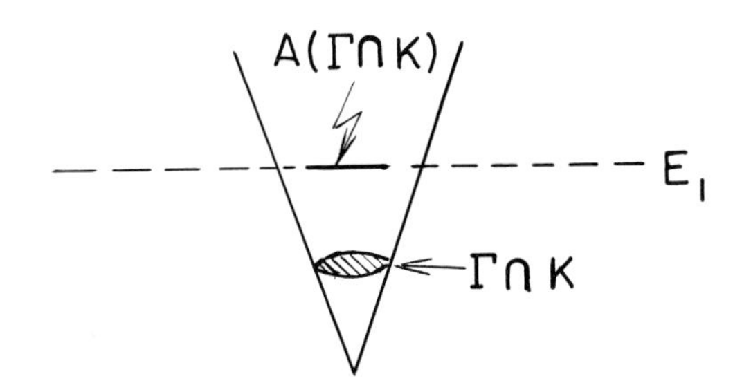

$\alpha_j(x)$, $x \in \partial G \cap K$, $j = 1,2,\ldots,k$. The functions $\alpha_j(x)$ can be extended as continuous functions $\hat{\alpha}_j(x)$, $x \in \overline{G}$. This gives an extension $\hat{A}$: $\hat{A}x = \sum \hat{\alpha}_j(x)\xi_j$ of A to $\overline{G}$ and $\hat{A}: \overline{G} \to E_1$.

In the relative topology of T, let u_0 be an interior point of T. Let P take E_1 onto T be the projection operator which is the identity on T and for $u \notin T$, Pu is the intersection of the line segment through u_0 and u with the boundary of T.

If $P\hat{A} = \tilde{A}$, then $\tilde{A}: \overline{G} \to T$ coincides with A on $\partial G \cap K$ and

$$\min_{\varphi \in \overline{G}} |\tilde{A}\varphi| > 0.$$

If A has no positive eigenvectors on $\partial G \cap K$, then $\tilde{A}$ has no eigenvectors on all of the boundary ∂G to which nonnegative eigenvalues correspond. Therefore, the vector fields

$$F(\varphi;\mu) = \varphi - \mu\tilde{A}\varphi, \quad F(\cdot;\mu): \partial G \to E_1$$

is nonzero on ∂G for $\mu \geq 0$. All of the $F(\varphi;\mu)$ are therefore homotopic on ∂G. Thus, the index $i(F(\cdot;\mu),\partial G) = i(F(\cdot;0),\partial G)$ is independent of μ.

Since $F(\cdot;0) = I$, $i(F(\cdot;0),\partial G)$ is 1. We now show that for any

$v_0 \in K$, $|v_0| = 1$, there is a $\mu_0 > 0$ such that all of the maps $F(\cdot;\mu)$, $\mu \geq \mu_0$, cannot take on the direction v_0. In fact, if no such μ_0 exists, then there is a sequence of $\varphi_n \in \partial G$, $\mu_n \geq 0$, $\mu_n \to \infty$ as $n \to \infty$, $n = 1,2,\ldots,$ such that

$$\frac{\varphi_n - \mu_n \tilde{A}\varphi_n}{|\varphi_n - \mu_n \tilde{A}\varphi_n|} = v_0 = \frac{\frac{\varphi_n}{\mu_n} - \tilde{A}\varphi_n}{\left|\frac{\varphi_n}{\mu_n} - \tilde{A}\varphi_n\right|}, \quad n = 1,2,\ldots .$$

Choose a subsequence of the φ_n and relabel the same so that $\varphi_n \to \varphi^*$ as $n \to \infty$. Taking the limit in the above expression yields $\tilde{A}\varphi^* = -|\tilde{A}\varphi^*| v_0$; that is, $\tilde{A}\varphi^* \notin T$, which is a contradiction. Since $F(\cdot;\mu)$, $\mu \geq \mu_0$, cannot take on the direction v_0, it follows that $i(F(\cdot,\mu),\partial G) = 0$. This contradiction proves the theorem for the finite dimensional case.

Now suppose E is an arbitrary Banach space. Assume there are no positive eigenvectors of A on $\partial G \cap K$. Then there is an $a > 0$ such that

$$|A\varphi - t\varphi| \geq 2a, \quad \varphi \in \partial G \cap K, \quad t \geq 0. \tag{27.3}$$

In fact, if this is not so, then there are sequences $\varphi_n \in \partial G \cap K$, $t_n \geq 0$, $n = 1,2,\ldots,$ such that

$$|A\varphi_n - t_n\varphi_n| < \frac{1}{n}, \quad n = 1,2,\ldots .$$

If

$$r_1 = \inf_{\varphi \in \partial G} |\varphi|, \qquad r_2 = \inf_{\varphi \in \partial G \cap K} |A\varphi|$$

$$R_1 = \sup_{\varphi \in \partial G} |\varphi|, \qquad R_2 = \sup_{\varphi \in \partial G \cap K} |A\varphi|$$

then r_1, r_2, R_1, R_2 are finite and positive. Furthermore, for $\varphi \in \partial G \cap K$, and $n \geq n_0$ sufficiently large,

$$t_n \leq \frac{|A\varphi_n| + |A\varphi_n - t_n\varphi_n|}{|\varphi_n|} < \frac{1}{r_1}(R_2 + \frac{1}{n})$$

$$t_n \geq \frac{|A\varphi_n| - |A\varphi_n - t_n\varphi_n|}{|\varphi_n|} > \frac{1}{|\varphi_n|}(r_2 - \frac{1}{n}) \geq \frac{1}{R_1}(r_2 - \frac{1}{n}).$$

It follows from these inequalities and the complete continuity of A that there are a subsequence which we again label as above, a $t_0 > 0$, $v_0 \in K$, such that $t_n \to t_0$, $A\varphi_n \to v_0$ as $n \to \infty$. This implies $\varphi_n \to \varphi_0 = v_0/t_0$ since

$$|\varphi_n - \varphi_0| \leq |\varphi_n - \frac{1}{t_n} A\varphi_n| + \frac{1}{t_n}|A\varphi_n - v_0| + |(\frac{1}{t_n} - \frac{1}{t_0})v_0|$$

$$\leq \frac{1}{nt_n} + \frac{1}{t_n}|A\varphi_n - v_0| + |\frac{1}{t_n} - \frac{1}{t_0}| \cdot |v_0|.$$

It is also clear that $A\varphi_0 = t_0\varphi_0$ which contradicts the assumption that A has no positive eigenvectors on $\partial G \cap K$. Therefore, (27.3) is satisfied.

We now approximate A on $\partial G \cap K$ by a finite dimensional (the range is finite dimensional) continuous operator A_1 such that

$$|A\varphi - A_1\varphi| < a, \quad \varphi \in \partial G \cap K.$$

This can always be done since A is completely continuous. In fact, choose a finite a-net in the compact set which is the closure B of $A(\partial G \cap K)$. Let this net consist of n points $u_1, \ldots, u_k$. Since A is positive, each $y_j \in K$. For any $u \in B$, let

$$Pu = \frac{\sum_{j=1}^{k} \mu_i(u) u_i}{\sum_{j=1}^{k} \mu_i(u)}$$

where

$$\mu_i(u) = \begin{cases} a - |u-u_i| & \text{if } |u-u_i| \leq a, \\ 0 & \text{if } |u-u_i| > a. \end{cases}$$

The continuous operator P takes every $u \in B$ into an element of the convex hull of those points of the a-net whose distance from u is less than or equal to a. Therefore $|Pu-u| < a$, $u \in B$. For any $\varphi \in \partial G \cap K$,

$$|PA\varphi - A\varphi| < a$$

$$|PA\varphi - t\varphi| \geq |A\varphi-\varphi| - |A\varphi-PA\varphi| > a.$$

Let $A_1 = PA\colon \partial G \cap K \to K$. The previous relation shows that A_1 has no positive eigenvectors on $\partial G \cap K$.

The range of A_1 lies in a finite dimensional subspace $E_1 \subset E$. Let $K_1 = K \cap E_1$, $G = G \cap E_1$. K_1 is a cone (or truncated cone) in E_1. Since $\partial G_1 \subset \partial G$, the operator A_1 is defined on $\partial G_1 \cap K_1$, $A_1\colon \partial G_1 \cap K_1 \to K_1$, and $\inf_{\varphi \in \partial G_1 \cap K_1} |A_1\varphi| > 0$. Therefore, A_1 has a positive eigenvector on $\partial G_1 \cap K_1$. This is a contradiction and proves the theorem.

Suppose K is a cone (or a truncated cone). Suppose $A\colon K \to K$ and let F be the set of positive eigenvectors of A. We say F forms a <u>continuous branch of length</u> r if for every open neighborhood G of zero with $\overline{G} \subset B_r \overset{\text{def}}{=} \{\varphi \in E\colon |\varphi| < r\}$, the boundary ∂G of G satisfies $\partial G \cap F \neq \emptyset$. We say F forms a <u>continuous branch of maximal length</u> if it forms a continuous branch of length r for every r for which $S_r \cap K \neq \emptyset$.

<u>Theorem 27.2.</u> Suppose A is positive with respect to K, is completely continuous and F is the set of positive eigenvectors of A. If A has a continuous branch of eigenvectors of maximal length and

(α) there exist numbers $M > 0$, $\mu^* > 0$ such that for all φ in F with $|\varphi| = M$, the associated eigenvalue $\mu < \mu^*$,

(β) there exists an open bounded neighborhood G of zero, $\overline{G} \subset B_M$, such that for each φ in $\partial G \cap F$, the associated eigenvalue $\mu > \mu^*$, then μ^* is an eigenvalue of A with associated eigenvector $\varphi^* \in K \cap (B_M \setminus \overline{G})$.

Proof. Assume μ^* is not an eigenvalue of A with associated eigenvector $\varphi^* \in F$. Let

$$F_0 = \{\varphi \in F\colon \varphi \in \overline{B}_M \setminus G\}$$

$$F_< = \{\varphi \in F_0\colon A\varphi = \mu\varphi,\ \mu < \mu^*\}$$

$$F_> = \{\varphi \in F_0\colon A\varphi = \mu\varphi,\ \mu > \mu^*\}.$$

Then $F_< \cap F_> = \emptyset$, $F_< \cup F_> = F_0$.

Since $F_< \subset F_0$, $F_> \subset F_0$ it follows that there is an $r > 0$ such that $\operatorname{dist}(F_<, 0) > r$, $\operatorname{dist}(F_>, 0) > r$. Furthermore, $F_> \subset B_M$.

The sets $F_<$, $F_>$ are closed. In fact, suppose $\varphi_n \in F_<$, $n = 1,2,\ldots$, $A\varphi_n = \mu_n\varphi_n$, $0 < \mu_n < \mu^*$ is a sequence converging to φ_0. Since $|\varphi_n| > r$, $n = 1,2,\ldots$, it follows that $\varphi_0 \neq 0$. Since $A\varphi_n \to A\varphi_0$ as $n \to \infty$, it follows that $\mu_n = |A\varphi_n|/|\varphi_n|$ converges as $n \to \infty$ to $\mu_0 = |A\varphi_0|/|\varphi_0|$. It follows that $\mu_0 < \mu^*$ for otherwise μ^* would be an eigenvalue of A with eigenvector $\varphi_0 \in F$. This shows $F_<$ is closed. To show $F_>$ is closed, suppose $\varphi_n \in F_>$, $A\varphi_n = \mu_n\varphi_n$, $\mu_n > \mu^*$, $n = 1,2,\ldots$, is a sequence converging to φ^0 as $n \to \infty$. As before, $\varphi_0 \neq 0$, $A\varphi_n \to A\varphi_0$ and $\mu_n \to \mu_0 = |A\varphi_0|/|\varphi_0|$ as $n \to \infty$. It follows that $\mu_0 > \mu^*$ which shows $F_>$ is closed.

We next show that $\operatorname{dist}(F_<, F_>) > 0$. If this is not so, there are two sequences $\{\varphi_n\} \subset F_<$, $\{\psi_n\} \subset F_>$, $A\psi_n = \mu_n\psi_n$, $\mu_n > \mu^*$, such that

$$\lim_{n \to \infty} |\varphi_n - \psi_n| = 0. \tag{27.4}$$

We know that $r < |\varphi_n|$, $r < |\psi_n| \leq M$, $n = 1,2,\ldots$. Since A is compact, there is a constant $N > 0$ such that $|A\psi_n| \leq N$, $n = 1,2,\ldots$. Therefore, $\mu^* < \mu_n \leq N/r$, $n = 1,2,\ldots$. Thus, there exist subsequences $\{\mu_{n_j}\} \subset \{\mu_n\}$, $\{A\psi_{n_j}\} \subset \{A\psi_n\}$ and

$\mu_0 \geq \mu^*$, ψ_0 such that $\mu_{n_j} \to \mu_0$, $A\psi_{n_j} \to \mu_0\psi_0$ as $j \to \infty$. Since $|A\psi_n| = \mu_n|\psi_n| \geq \mu^* r$, it follows that $\psi_0 \neq 0$. Furthermore,

$$|\psi_{n_j} - \psi_0| = \left|\frac{A\psi_{n_j}}{\mu_{n_j}} - \psi_0\right| = \left|\frac{A\psi_{n_j} - \mu_0\psi_0}{\mu_{n_j}}\right| + \left|(\frac{\mu_0}{\mu_{n_j}} - 1)\psi_0\right|$$

$$\leq \frac{1}{\mu^*}|A\psi_{n_j} - \mu_0\psi_0| + \left|\frac{\mu_0}{\mu_{n_j}} - 1\right| \cdot |\psi_0|.$$

This shows that $\psi_{n_j} \to \psi_0$ as $n \to \infty$. Furthermore, $F_>$ closed implies $\psi_0 \in F_>$. On the other hand, (27.4) implies φ_{n_j} also approaches ψ_0 and closure of $F_<$ implies $\psi_0 \in F_<$. But, this contradicts the fact that $F_< \cap F_> = \emptyset$. Thus,

$$\text{dist}(F_<, F_>) = d > 0.$$

If we define

$$G^* = G \cup G_>, \quad G_> = \{g \in E: |\varphi - g| < d/2, \ \varphi \in F_>\},$$

then G^* is a bounded, open neighborhood of zero. The set $\partial G^* \cap K$ is exterior to $G \cap K$. Therefore, any eigenvectors of A in $\partial G^* \cap K$ are in F_0. On the other hand, the construction of G^* implies that no such eigenvectors can be in either $F_>$ or $F_<$. This shows $\partial G^* \cap K$ has no eigenvector which is a contradiction to the basic assumption of a continuous branch of eigenvectors of maximal length. This proves the theorem.

<u>Corollary 27.1.</u> If the conditions of Theorem 27.2 are satisfied for $\mu^* = 1$, then A has a fixed point in $K \cap (B_M \setminus G^*)$.

An immediate consequence of Theorem 27.1 and 27.2 is

<u>Theorem 27.3.</u> Suppose K is a cone (or a truncated cone), A is positive with respect to K, is completely continuous and F is the set of positive eigenvectors of A. If

(α) for any open set $G \subset E$, $0 \in G$,

$$\inf_{\varphi \in \partial G \cap K} |A\varphi| > 0;$$

(β) there exists an $M > 0$ such that $\varphi \in F$, $|\varphi| = M$, $A\varphi = \mu\varphi$ implies $\mu < 1$;

(γ) there exists an open neighborhood H of zero, $\overline{H} \subset B_M$, such that $\varphi \in \partial H \cap F$, $A\varphi = \mu\varphi$ implies $\mu > 1$;

then A has a fixed point in $K \cap (B_M \setminus \overline{H})$.

In the next section, we also need the following lemma.

Lemma 27.3. If K is a cone or a truncated cone, P is a projection operator, and for every $\delta > 0$ for which $\partial B(\delta) \cap K \neq \emptyset$, $a(\delta) \overset{\text{def}}{=} \inf_{\varphi \in \partial B(\delta) \cap K} |P\varphi| > 0$, then there is a $\nu > 0$ such that $a(\delta) \geq \nu\delta$. Furthermore, there is a $\delta_0 > 0$ such that $a(\delta) = \nu\delta$ for $0 < \delta \leq \delta_0$.

Proof. It is no loss in generality to assume that K is solid. There is a $\delta_1 > 0$ such that $\varphi \in (\overline{B}(\delta_1) \cap K) \cap (\partial K)$ implies $\lambda\varphi \in \partial K$, $0 \leq \lambda \leq 1$, and $\varphi \in (\partial B(\delta_1) \cap K)$, implies $\varphi \in$ interior of K. Fix such a δ_1. For any $\delta > 0$ for which $\partial B(\delta) \cap K \neq \emptyset$,

$$\inf_{\varphi \in \partial B(\delta) \cap K} |P\varphi| = \inf_{\varphi \in \partial B(\delta) \cap K} \left| P\left(\frac{\varphi\delta_1}{|\varphi|}\right) \right| \frac{|\varphi|}{\delta_1}$$

$$\geq \inf_{\psi \in \partial B(\delta_1) \cap K} |P\psi| \frac{\delta}{\delta_1} = \frac{a(\delta_1)}{\delta_1} \delta.$$

If $\nu = a(\delta_1)/\delta_1$, then the first statement of the lemma is true. Also, if $\delta \leq \delta_1$, the inequalities in the estimates become equalities and the last conclusion of the lemma holds with $\delta_1 = \delta_0$.

28. A PERIODICITY THEOREM FOR FUNCTIONAL EQUATIONS

Consider the linear system

$$\dot{u}(t) = L(u_t). \tag{28.1}$$

Let the generalized eigenspaces of (28.1) corresponding to the eigenvalues with positive real parts generate a linear subspace U of C. Decompose the space C as

$$C = U \oplus S$$

as in Section 26. Let π_U, $\pi_S = I - \pi_U$ be the corresponding projection operators onto U, S, respectively. Along with (28.1), we consider the nonlinear system

$$\dot{x}(t) = L(x_t) + f(x_t) \tag{28.2}$$

where

$$\begin{aligned} &f(0) = 0 \\ &|f(\varphi) - f(\psi)| \leq \mu(\sigma)|\varphi-\psi| \end{aligned} \tag{28.3}$$

for $|\varphi|$, $|\psi| \leq \sigma$ and $\mu(\sigma)$ is continuous and nondecreasing with $\mu(0) = 0$.

Suppose K is a cone (or a truncated cone) such that for any $\varphi \in K$, there is a time $\tau(\varphi) > 0$ such that the solution $x(\varphi)$ of (28.2) with initial value φ at zero satisfies $x_{\tau(\varphi)}(\varphi) \in K$. If we let $A\varphi = x_{\tau(\varphi)}(\varphi)$, $\varphi \in K$, then $A: K \to K$ is a positive operator.

Theorem 28.1. Suppose A is the same as defined above, $\tau: K \to [0,\infty)$ is continuous, $\tau(\varphi) \geq r$, $\varphi \in K$, τ and A take closed bounded sets into bounded sets and the following conditions are satisfied:

(I) For any open bounded set $G \subset E$, $0 \in G$,

$$\inf_{\varphi \in \partial G \cap K} |A\varphi| > 0.$$

(II) If F is the set of positive eigenvectors of A, there is an $M > 0$ such that $\varphi \in F$, $|\varphi| = M$, $A\varphi = \mu\varphi$ implies $\mu < 1$.

(III) For any $\delta > 0$,

$$\inf_{\varphi \in \partial B(\delta) \cap K} |\pi_U \varphi| > 0.$$

Under these conditions, there exists a nontrivial periodic solution of (28.2) with period greater than r.

Proof. Since $x_t(\varphi)$ is continuous in t, φ and $\tau(\varphi)$ is continuous, the operator A is continuous. For any bounded set B in K, AB is bounded. Therefore, there is a constant $m > 0$ such that $|x_t(\varphi)| \leq m$, $0 \leq t \leq \tau(\varphi)$, $\varphi \in B$. Thus, there is an $N > 0$ such that $|f(t, x_t(\varphi))| \leq N$ for $0 \leq t \leq \tau(\varphi)$, $\varphi \in B$. Since $\tau(\varphi) \geq r$, this implies the set AB is relatively compact. Consequently, A is completely continuous. Since hypotheses (I) and (II) are the same as (α), (β) in Theorem 27.3, it remains only to show that (γ) of that theorem is satisfied since a fixed point of A obviously corresponds to a periodic solution of (28.2).

For any $\delta_0 > 0$, there is a $\delta_1 > 0$ so that $|A\varphi| \leq \delta_0$ if $|\varphi| \leq \delta \leq \delta_1$. In fact, for any $\delta > 0$, there is an $m(\delta) > 0$ so that $r \leq \tau(\varphi) \leq m(\delta)$, $\varphi \in B(\delta) \cap K$. Let $m_1 = m_1(\delta_0) = \sup_{0 \leq \delta_1 \leq \delta_0} m(\delta)$. Since $x_t(0) = 0$, $x_t(\varphi)$ is continuous in t, φ and m_1 is finite, the existence of δ_1 follows immediately.

Choose δ_0 so that the conclusion $a(\delta) = \nu\delta$ of Lemma 27.3 is true. For $p < \nu$, $0 < p < 1$, restrict δ_0 further so that the conclusion of Theorem 27.2 is true. Then $\pi_U x_t(\varphi)$, $\varphi \in K$, is increasing in t as long as $\pi_U x_t(\varphi) \geq p\delta$, $|x_t(\varphi)| \leq \delta$ for $0 < \delta \leq \delta_1$. If, for any $\delta > 0$,

$$G_\delta = \{\varphi \in C, |\varphi| < \delta\} \cap \{\varphi \in C: |\pi_U\varphi| < p\delta\}$$

then G_δ is a bounded open set, $0 \in G$ and

$$\partial G_\delta \cap K = \{\varphi \in K: |\pi_U\varphi| = p\delta\}.$$

From the above construction, it follows that any $\varphi \in \partial G_\delta \cap K$, $A\varphi = \mu\varphi$, $0 < \delta \leq \delta_1$, has $\mu > 1$. Therefore, (γ) of Theorem 27.3 is satisfied and the theorem is proved.

Condition I of Theorem 28.1 will be satisfied if one shows that

(I') There is an $M > 0$ such that $|Ax| < |x|$ for all $|x| > M$.

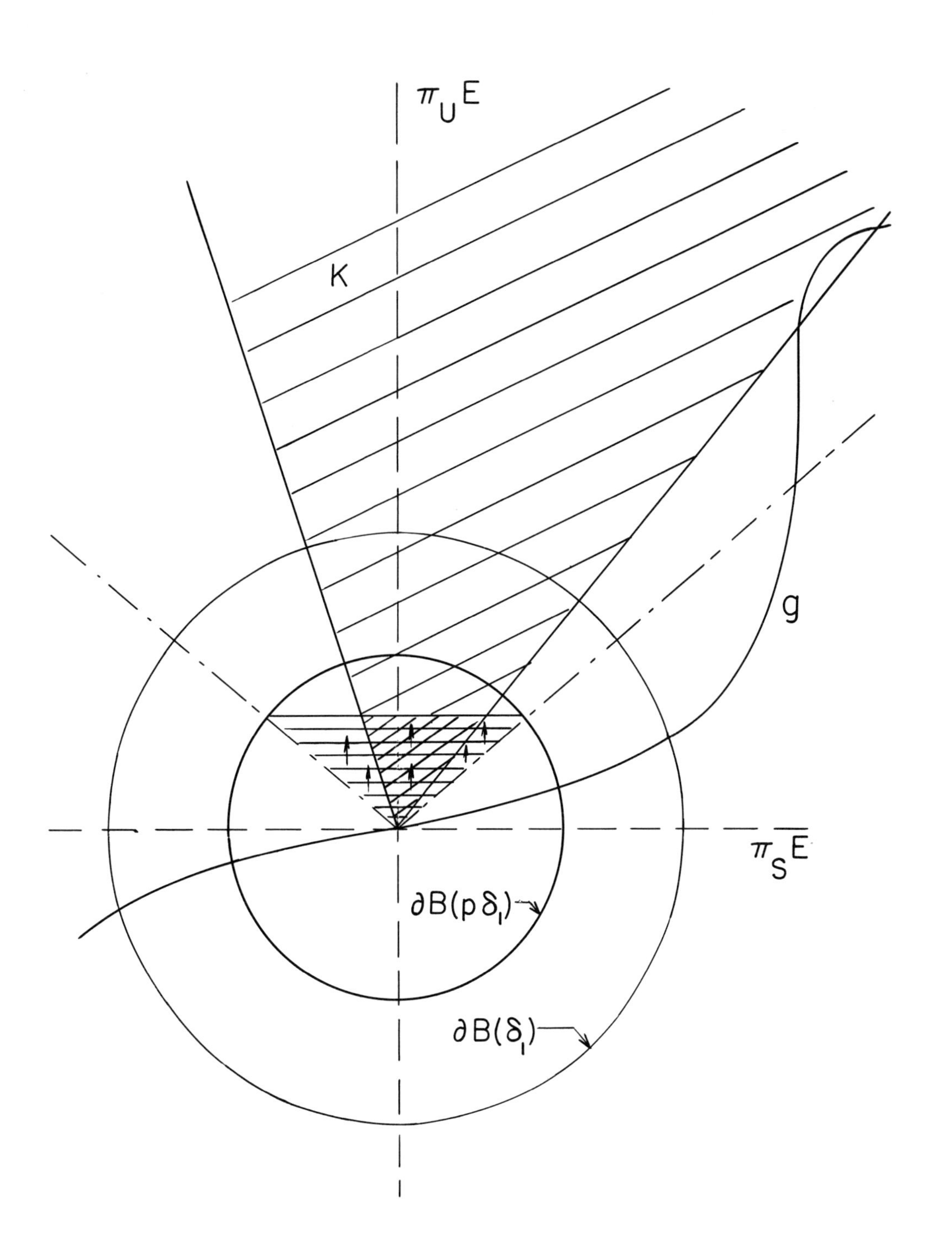

$\pi_U E$
K
g
$\pi_S E$
$\partial B(p\delta_1)$
$\partial B(\delta_1)$

29. THE EQUATION $\dot{x}(t) = -\alpha x(t-1)[1+x(t)]$

Consider the equation

$$\dot{x}(t) = -\alpha x(t-1)[1+x(t)] \tag{29.1}$$

where $\alpha > 0$. It is clear that $x(\varphi)(t) > -1$, $t \geq 0$, if $\varphi(0) > -1$. Also, it is clear that there is no $t_0 > 0$ such that $x(\varphi)(t) = 0$ for $t \geq t_0$ unless $\varphi = 0$. We say the zeros of $x(\varphi)$ are bounded if $x(t)$ has only a finite number of positive zeros.

<u>Lemma 29.1</u>. (i) If $\varphi(0) > -1$ and the zeros of $x(\varphi)$ are bounded then $x(\varphi)(t) \to 0$ as $t \to \infty$.

(ii) If $\varphi(0) > -1$, then $x(\varphi)(t)$ is bounded. Furthermore, if the zeros of $x(\varphi)$ are unbounded, then any maximum of $x(\varphi)(t)$, $t > 0$, is less than $e^{\alpha}-1$.

(iii) If $\varphi(0) > -1$ and $\alpha > 1$, then the zeros of $x(\varphi)$ are unbounded.

(iv) If $\varphi(\theta) > 0$, $-1 < \theta < 0$ [or if $\varphi(0) > -1$, $\varphi(\theta) < 0$, $-1 < \theta < 0$], then the zeros (if any) of $x(\varphi)(t)$ are simple and the distance from a zero of $x(\varphi)(t)$ to the next maximum or minimum is ≥ 1.

<u>Proof</u>. (i) Suppose there is a $t_1 > 0$ such that $x(t) = x(\varphi)(t)$ is of constant sign for $t \geq t_1 - 1$. Since $x(\varphi)(t) > -1$ for all $t \geq 0$, $\dot{x}(t)x(t-1) < 0$, $t \geq t_1$. Therefore, $x(t)$ is bounded and approaches a limit monotonically. This implies $\dot{x}(t)$ is bounded and therefore $\dot{x}(t) \to 0$ as $t \to \infty$. This implies $x(t) \to 0$ or -1, but -1 is obviously ruled out.

(ii) $x = x(\varphi)$ satisfies

$$1 + x(t) = [1 + x(t_0)]e^{-\alpha \int_{t_0-1}^{t-1} x(\xi)d\xi} \tag{29.2}$$

for any $t \geq t_0 \geq 0$. If the zeros of $x(t)$ are bounded, then (i) implies x is bounded. If there is a sequence of nonoverlapping intervals I_k of $[0,\infty)$

such that x is zero at the endpoints of each I_k and has constant sign on I_k, then there is a t_k such that $\dot{x}(t_k) = 0$. Thus, $x(t_k-1) = 0$. Consequently, (29.2) implies for $t_0 = t_k-1$, $t = t_k$,

$$\ln(1+x(t_k)) = -\alpha \int_{t_k-2}^{t_k-1} x(\xi)d\xi < \alpha$$

since $x(t) > -1$, $t \geq 0$. Finally $x(t_k) \leq e^{\alpha}-1$ for all t_k. This proves (ii).

(iii) If the zeros of $x = x(\varphi)$ are bounded, then (i) implies $x(t) \to 0$ as $t \to \infty$ and, thus, the existence of a $t_0 > 0$ such that $\alpha(1+x(t)) > 1$ for $t \geq t_0$ and $x(t)$ has constant sign for $t \geq t_0$. Thus,

$$\begin{aligned}\dot{x}(t)x(t-1) &= -\alpha x^2(t-1)(1+x(t)) \\ &< -x^2(t-1) < 0, \quad t \geq t_0+1,\end{aligned}$$

and $x(t) \to 0$ monotonically as $t \to \infty$. If x is positive on $[t_0,\infty)$, then

$$\begin{aligned}x(t_0+3) - x(t_0+2) &= \int_{t_0+2}^{t_0+3} \dot{x}(t)dt \\ &< -\int_{t_0+1}^{t_0+2} x(t)dt < -x(t_0+2)\end{aligned}$$

and $x(t_0+3) < 0$. This is a contradiction. If $x(t)$ is negative on $[t_0,\infty)$, then a similar contradiction is obtained.

(iv) Suppose $x(t_0) = 0$ and $x(t) > 0$, $t_0-1 < t < t_0$. For $t_0 < t < t_0+1$, $\dot{x}(t) < 0$. Similarly, if $x(t) < 0$ for $t_0-1 < t < t_0$ and $x(t_0) = 0$, then $\dot{x}(t) > 0$, $t_0 < t < t_0+1$. Therefore, the assertion of (iv) is obvious.

Note that $\alpha > \pi/2$ and $\varphi(\theta) > 0$, $-1 < \theta < 0$, implies the zeros of $x(\varphi)$ are unbounded and the distance from a zero to the next maximum or minimum is > 1.

Set K as the class of all functions $\varphi \in C$ such that $\varphi(\theta) \geq 0$, $-1 < \theta \leq 0$, $\varphi(-1) = 0$, φ nondecreasing. Then K is a cone. If $\alpha > 1$, $\varphi \in K$,

$\varphi \neq 0$, let $z(\varphi) = \min\{t: x(\varphi)(t) = 0, \dot{x}(\varphi)(t) > 0\}$. This minimum exists from Lemma 29.1iii) and iv). Also $z(\varphi) > 2$. Furthermore, Lemma 29.1iv) implies $x(\varphi)(t)$ is positive and nondecreasing on $(z(\varphi), z(\varphi)+1)$. Consequently, if $\tau(\varphi) = z(\varphi)+1$, then the mapping

$$A\varphi = x_{\tau(\varphi)}(\varphi),$$

is a positive mapping relative to K. Since $\dot{x}(\varphi)(\tau(\varphi)-1) > 0$, continuity of $x(\varphi)(t)$ in t, φ implies that $\tau(\varphi)$ is continuous in K. Define $\tau(0) = \lim_{\varphi \to 0} \tau(\varphi)$. From Lemma 29.1ii) and iv) $|A\varphi| \leq e^{\alpha}-1$ and thus A takes bounded sets into bounded sets.

Lemma 29.2. If $\alpha > 1$, $\tau: K \to [0,\infty)$ takes closed bounded sets into bounded sets.

Proof. For any closed bounded set in K, $0 \leq \varphi(\theta) \leq \beta$, it follows from (29.1) that $\dot{x} \leq 0$ on $(0,1)$ and thus $-1 \leq x(\varphi)(t) \leq \beta$, $0 \leq t \leq 1$. Consequently, $\dot{x}(\varphi)(t)$ is bounded on $[0,1]$. This implies the set $\{x_1(\varphi), 0 \leq \varphi(\theta) \leq \beta\}$ is relatively compact. Now $z(\varphi) = \tau(\varphi)-1$ satisfies

$$z(\varphi) = \min\{t: x(\varphi)(t) = x(x_1(\varphi))(t-1) = 0, \dot{x}(x_1(\varphi))(t-1) > 0\}.$$

Thus $z(\varphi)$ can be considered a function of $x_1(\varphi)$ and it is continuous in this variable. This shows $z(\varphi)$ is bounded for $0 < \varphi(\theta) \leq \beta$.

Lemma 29.3. If $\alpha > 1$, G is an open bounded neighborhood of zero, then

$$\inf_{\varphi \in \partial G \cap K} |A\varphi| > 0.$$

Proof. If $\inf_{\varphi \in \partial G \cap K} |A\varphi| = 0$, then there is a sequence $\varphi_n \in \partial G \cap K$ such that $|A\varphi_n| \to 0$. We may choose a subsequence of the φ_n so that $\tau(\varphi_n) \to \tau_0 \geq 1$ as $n \to \infty$. Therefore, $x(\varphi_n)(t) \to y(t)$ as $n \to \infty$ uniformly for $t \in [0, \tau_0]$ and $y(t)$ must correspond to a solution of the equation on $[1, \tau_0]$. It is clear from (29.1) that $y(t) = 0$, $0 \le t \le \tau_0$. Consequently $\varphi_n(0) \to 0$ as $n \to \infty$ and the monotonicity of the φ_n imply that $\varphi_n \to 0$ as $n \to \infty$. But this is impossible, since there is a $\beta > 0$ such that $|\varphi_n| \ge \beta$.

<u>Lemma 29.4.</u> If $\alpha > \pi/2$ there is a zero $\lambda = \gamma + i\sigma$ of

$$(29.3) \qquad \lambda e^{\lambda} = -\alpha$$

with $\gamma > 0$, $0 < \sigma < \pi$.

<u>Proof.</u> The function $\rho(\mu) = -\mu e^{\mu}$ satisfies $\rho'(\mu) = -(1+\mu)e^{\mu}$ and, therefore, $\rho'(\mu) > 0$, $-\infty < \mu < -1$, $\rho'(-1) = 0$, $\rho'(\mu) < 0$, $-1 < \mu < \infty$. Consequently, $\rho(\mu)$ is a maximum at $\mu = -1$, $\rho(-1) = e^{-1}$. Therefore, the equation (29.3) has no real roots for $\alpha > e^{-1}$. If $\alpha > e^{-1}$, $\lambda = \gamma + i\sigma$, $\mu = -\gamma$, satisfies (29.3), then $\mu - i\sigma = \alpha \exp(\mu - i\sigma)$ and

$$\mu = \alpha e^{\mu} \cos \sigma, \quad \sigma = \alpha e^{\mu} \sin \sigma$$

or

$$\mu = \sigma \cot \sigma, \quad \alpha = \frac{\sigma e^{-\sigma \cot \sigma}}{\sin \sigma} \overset{\text{def}}{=} f(\sigma).$$

Let us consider $f(\sigma)$ for $0 < \sigma < \pi$. It is clear that $f(\sigma) > 0$.

$$\frac{f'(\sigma)}{f(\sigma)} = \frac{1}{\sigma} - 2 \cot \sigma + \sigma \operatorname{cosec}^2 \sigma$$

$$= \frac{(1-\sigma \cot \sigma)^2 + \sigma^2}{\sigma} > 0.$$

Furthermore, $f(\sigma) \to \infty$ as $\sigma \to \pi$, $f(\sigma) \to e^{-1}$ as $\sigma \to 0$. Therefore, there is exactly one value of σ say $\sigma_0 = \sigma_0(\alpha)$, $0 < \sigma_0(\alpha) < \pi$, for which $f(\sigma_0(\alpha)) = \alpha$ if $\alpha > e^{-1}$. Let $\gamma_0 = -\sigma_0 \cot \sigma_0$. Note that $f(\pi/2) = \pi/2$. Therefore, $\gamma_0 > 0$

if $\alpha > \pi/2$. This proves the lemma.

Decompose C as in Theorem 28.1.

<u>Lemma 29.5.</u>

$$\inf_{\varphi \in \partial B(1) \cap K} |\pi_U \varphi| > 0$$

if $\alpha > \pi/2$.

<u>Proof.</u> Let λ_0, $R(\lambda_0) > 0$, be the eigenvalue assured by Lemma 29.3. Let $\varphi = e^{\lambda_0 \theta}/(1+\lambda_0)$, $-1 \leq \theta \leq 0$, $\psi(s) = e^{-\lambda_0 s}$, $0 \leq s \leq 1$, $\Phi = (\varphi, \bar{\varphi})$, $\Psi = (\psi, \bar{\psi})$. The adjoint for the linear part of (29.1) is

$$\dot{y}(t) = \alpha y(t+1)$$

and the bilinear form is

$$(\psi, \varphi) = \psi(0)\varphi(0) + \alpha \int_{-1}^{0} \psi(\xi+1)\varphi(\xi)d\xi.$$

It is easily seen that (Ψ, Φ) = the identity.

If there is a sequence $\{\varphi_n\}$ in $\partial B(1) \cap K$ such that $\pi_U \varphi_n \to 0$ as $n \to \infty$, then necessarily $|(\Psi, \varphi_n)| \to 0$ as $n \to \infty$. If we let R_n, I_n be the real and imaginary parts respectively of (Ψ, φ_n), then (since $\varphi_n(0) = 1$)

$$R_n = 1 - \alpha \int_{-1}^{0} \varphi_n(s) e^{-\gamma_0(s+1)} \cos \sigma_0(s+1) ds$$

$$I_n = \alpha \int_{-1}^{0} \varphi_n(s) e^{-\gamma_0(s+1)} \sin \sigma_0(s+1) ds$$

$$= \alpha \int_{0}^{1} \varphi_n(s-1) e^{-\gamma_0 s} \sin \sigma_0 s \; ds.$$

Since $0 < \sigma_0 < \pi$, and $I_n \to 0$ as $n \to \infty$, it follows that $\varphi_n(\theta) \to 0$, $-1 \leq \theta \leq 0$. Thus, $R_n \to 1$ as $n \to \infty$. This is a contradiction.

For $M > e^{\alpha}-1$, all of the conditions of Theorem 28.1 are satisfied and we have

<u>Theorem 29.1</u>. If $\alpha > \pi/2$, equation (29.1) has a nontrivial periodic solution.

30. THE EQUATION $\dot{x}(t) = -\alpha x(t-1)[1-x^2(t)]$

Consider the equation

$$\dot{x}(t) = -\alpha x(t-1)[1-x^2(t)] \tag{30.1}$$

where $\alpha > 0$. It is clear that $-1 < x(\varphi)(t) < 1$ if $-1 < \varphi(0) < 1$. Also, it is clear there is a $t_0 > 0$ such that $x(\varphi)(t) = 0$ for $t \geq t_0$ only if $\varphi = 0$.

Lemma 30.1. (i) If $-1 < \varphi(0) < 1$ and the zeros of $x(\varphi)(t)$ are bounded, then $x(\varphi)(t) \to 0$ as $t \to \infty$.

(ii) If $-1 < \varphi(0) < 1$, then $-1 \leq x(\varphi)(t) \leq 1$, $t \geq 0$, and if the zeros of $x(\varphi)$ are unbounded then any maximum [or minimum] of $x(t)$, $t > 0$, is less than $(e^{2\alpha}-1)/(e^{2\alpha}+1)$ [greater than $(e^{2\alpha}-1)/(e^{2\alpha}+1)$].

(iii) If $-1 < \varphi(0) < -1$ and $\alpha > 1$, then the zeros of $x(\varphi)$ are unbounded.

(iv) If $1 > \varphi(\theta) > 0$, $-1 < \theta < 0$ [or if $\varphi(0) > -1$, $\varphi(\theta) < 0$, $-1 < \theta < 0$], then the zeros (if any) of $x(\varphi)(t)$ are simple and the distance from a zero of $x(\varphi)(t)$ to the next maximum or minimum is ≥ 1.

Proof. The proof of (i), (iii), (iv) are the same as the proof in Lemma 29.1 except for obvious modifications. To prove (ii), observe that $x = x(\varphi)$ satisfies

$$\ell n \frac{1+x(t)}{1-x(t)} - \ell n \frac{1+x(t_0)}{1-x(t_0)} = -2\alpha \int_{t_0-1}^{t-1} x(\xi)d\xi$$

for any $t \geq t_0 \geq 1$. Using the same argument as in Lemma 29.1(iii) for this equation, one proves (ii) above.

Let $K = \{\varphi \in C: \varphi(-1) = 0, 0 \leq \varphi(\theta) < 1, -1 \leq \theta \leq 0, \varphi$ nondecreasing$\}$. Then K is a truncated cone. Define the operator $A: K \to K$ as in Section 29. Then A takes closed bounded sets into bounded sets and one proves τ does the same exactly as in the proof of Lemma 29.2. The analogue of Lemma 29.3 is proved for this case exactly the same as before. Lemma 29.4 and 29.5 are true for K.

From Lemma 30.1(ii), $|A\varphi| \leq (e^{2\alpha}-1)/(e^{2\alpha}+1) \overset{\text{def}}{=} \beta < 1$ for all $\varphi \in K$. If we choose $M > \beta$, then condition (II) of Theorem 28.1 is satisfied. Therefore, we have

<u>Theorem 30.1</u>. If $\alpha > \pi/2$, equation (30.1) has a nontrivial periodic solution.

31. THE EQUATION $\ddot{x}(t) + f(x(t))\dot{x}(t) + g(x(t-r)) = 0$

Consider the equation

$$\ddot{x}(t) + f(x(t))\dot{x}(t) + g(x(t-r)) = 0 \tag{31.1}$$

where $r > 0$, $f(x)$ is continuous, $g(x)$ has continuous first derivatives,

(a) $F(x) = \int_0^x f(s)ds$ is odd in x.

(b) $F(x) \to \infty$ as $|x| \to \infty$ and there is a $\beta > 0$ such that $F(x) > 0$ and is monotone increasing for $x > \beta$.

(c) $g'(x) > 0$, $xg(x) > 0$, $x \neq 0$, $g(x) = -g(-x)$, $g'(0) = 1$.

(d) $F^{-1}(x)g(F^{-1}(x))/x \to 0$ as $x \to \infty$.

Equation (31.1) is equivalent to the system,

$$\begin{aligned} \dot{x}(t) &= y(t) - F(x(t)), \\ \dot{y}(t) &= -g(x(t-r)). \end{aligned} \tag{31.2}$$

Let $z = (x,y)$, $\psi = (\varphi,a)$, where $\varphi \in C([-r,0],R^1)$, $a \in R^1$, $C_0 = C([-r,0],R^1) \times R^1$. For any $\psi \in C_0$, the initial value problem for (31.2) has a unique solution.

Let $K_0 = \{\psi = (\varphi,a) \in C_0\colon 0 \leq a < \infty,\ 0 = \varphi(-r) \leq \varphi(\theta),\ -r < \theta \leq 0$ and $\varphi(\theta)$ is nondecreasing in $\theta\}$. Then K_0 is a cone.

<u>Lemma 31.1.</u> (i) For any $\psi \in K_0$, the solution $z(\psi)$, of (31.2) is oscillatory; that is, both $x(\psi)(t)$ and $y(\psi)(t)$ have infinitely many zeros.

(ii) For any ψ in K_0, there is a finite time $\tau_1(\psi) > r$ such that the solution $z = z(\psi)$ of (31.2) satisfies

$$z_{\tau_1(\psi)} = (x_{\tau_1(\psi)}, y(\tau_1(\psi))) \in -K = \{-\psi\colon \psi \in K_0\}.$$

(iii) For any ψ in $-K_0$, there is a finite time $\tau_2(\psi) > r$ such that the solution $z = z(\psi)$ of (31.2) satisfies $z_{\tau_2(\psi)} \in K_0$.

<u>Proof</u>. If $\psi \in K$, $z(\psi)$ a solution of (31.2) implies $-z(\psi) = z(-\psi)$ is also a solution. This symmetry and (ii) implies (iii). Also (ii) and (iii) imply (i). Therefore, it is only necessary to prove (ii).

For any ψ in C_0, let $P\psi = \psi(0)$. If z_t is a solution of (31.2), then Pz_t traces out a curve in the (x,y)-plane. For $\psi \in K$, we analyze Pz_t where $z = z(\psi)$ is the solution of (31.2). In the accompanying figure, we have shown a typical Pz_t where $P\psi$ lies above the curve Γ: $y = F(x)$. Of course, we must prove that Pz_t has the form shown, but for notational purposes the point 0 corresponds to $P\psi$, the point 1 corresponds to the time t_1 at which Pz_{t_1} lies on Γ, the point 2 corresponds to the time t_2 at which Pz_{t_2} lies on the x axis, the point 3 corresponds to the time t_3 at which Pz_{t_3} lies on the y-axis and the point 4 corresponds to Pz_{t_4}, $t_4 = t_3+r$. If Pz_t crosses the curve as shown, then $0 < t_1 < t_2 < t_3 < t_4$ and if it crosses the x-axis between 0 and β, then $0 \leq t_2 < t_1 < t_3 < t_4$. Notice that PK_0 lies in the first quadrant.

Suppose $P\psi$ is above Γ. As long as Pz_t is above Γ, $\dot{x}(t) = y(t) - F(x(t)) > 0$, $\dot{y}(t) = -g(x(t-r)) \leq 0$. There must be a $t_1 > 0$ such that $Pz_t \in \Gamma$. If this were not so, then $x(t) \geq x(r) > 0$ for all $t \geq r$ and $\dot{y}(t) = -g(x(t-r)) \leq -x(r) < 0$ for $t \geq 2r$. Therefore, Pz_t must cross Γ.

As long as Pz_t is below Γ and $x(t) \geq 0$, $\dot{x}(t) < 0$, $\dot{y}(t) \leq 0$. The curve Pz_t cannot cross Γ in this region since Pz_t must cross Γ with a vertical slope. Then, $\dot{x}(t) < -\delta < 0$ in this region and there must exist a t_3 such that Pz_{t_3} is on the y-axis.

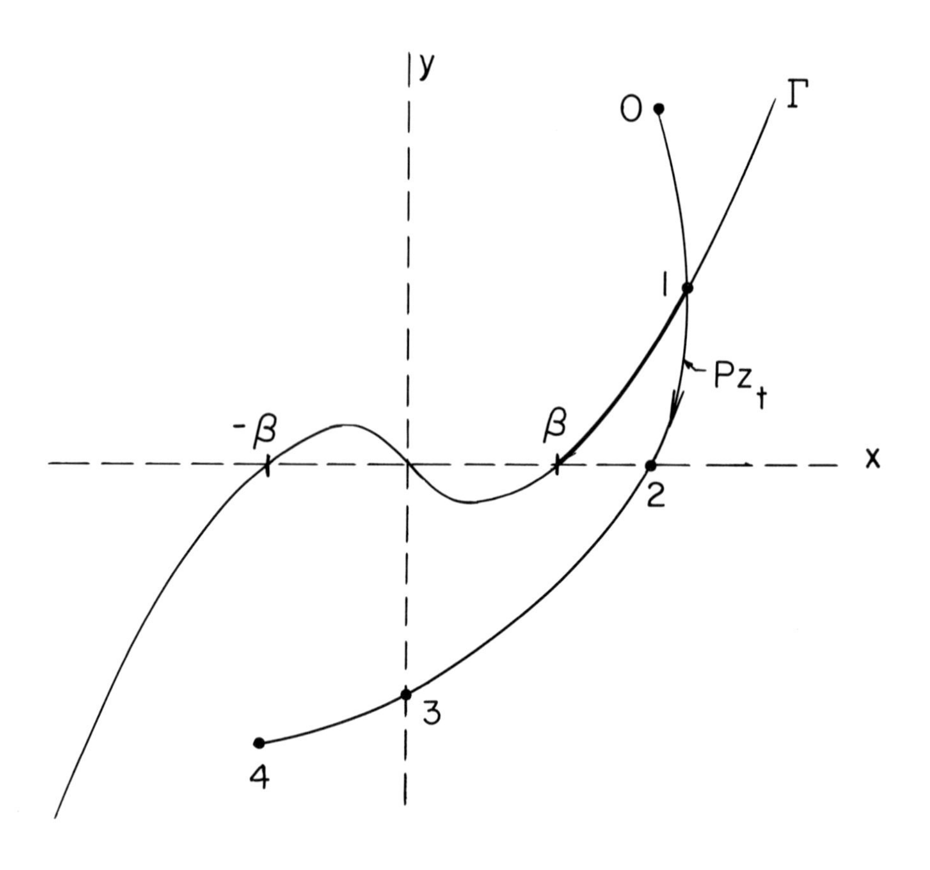

Furthermore, the curve Pz_t must cross Γ at some time $t_5 > t_3$ and the slope of the curve at this point is ∞. This means that $\dot{y}(t_5) > 0$ and thus, $x(t_5-r) < 0$. Therefore, $t_5 > t_3+r$ and the point 4 is situated as shown on the figure. If $\tau_1(\psi) = t_4$, then $z_{\tau_1(\psi)} \in -K_0$.

If $P\psi$ is between Γ and the y-axis, the same argument applies and proves the lemma.

For any $\psi \in K_0$, let $\tau_1(\psi)$ be the number given by Lemma 31.1(ii) and define $A: K_0 \to K_0$ by $A\psi = -z_{\tau_1(\psi)}(\psi)$. If $\psi \neq 0$, $A\psi = \psi$, then the symmetry in (31.2) implies that

$$z_{2\tau_1(\psi)}(\psi) = z_{\tau_1(\psi)}(z_{\tau_1(\psi)}\psi) = z_{\tau_1(\psi)}(-\psi) = -z_{\tau_1(\psi)}(\psi) = \psi$$

and ψ corresponds to a nontrivial periodic solution of (31.2) of period $2\tau_1(\psi)$.

Using the arguments similar to the proof of Lemma 29.2, one shows that $\tau_1 \colon K_0 \to [r,\infty)$ is continuous and maps bounded sets into bounded sets. The proof that

$$\inf_{\psi \in \partial G \cap K_0} |A\psi| > 0$$

for each open bounded neighborhood of zero is essentially the same as the proof of Lemma 29.3. It is not difficult to show directly from (31.2) that A takes bounded sets into bounded sets.

<u>Lemma 31.2</u>. There are $m_0 > 0$, $\delta > 0$ such that $\inf_{t\in[t_2(\psi),t_3(\psi)]}[-\dot{x}(\psi)(t)] \geq \delta > 0$ for all $\psi \in K_0$, $|\psi| = m \geq m_0$ where $t_2(\psi)$, $t_3(\psi) \geq 0$ are the first times for which $Pz_{t_2(\psi)}(\psi)$ is on the x-axis and $Pz_{t_3(\psi)}(\psi)$ is on the y-axis.

<u>Proof</u>. We first show that there is an $m_0 > 0$ such that $t_1(\psi) < t_2(\psi)$ for all ψ with $|\psi| = m \geq m_0$ if $P\psi$ is above Γ. Since $\psi = (\varphi,a)$, and $\varphi(\theta)$ is non-decreasing, it follows that $Pz_t(\psi)$ is above the curve defined by

$$\dot{x}(t) = y(t) - F(x(t))$$
$$\dot{y}(t) = -g(x(t))$$

as long as $Pz_t(\psi)$ is above Γ. But is is known that this latter curve must cross Γ with the x-coordinate greater than β if m is sufficiently large.

Therefore, for an appropriate m_0, we may assume $Pz_t(\psi)$, $|\psi| = m \geq m_0$ has the form shown in the figure. Suppose there are sequences $\psi_n \in K_0$, $|\psi_n| = m$, $t_n \in [t_2(\psi_n), t_3(\psi_n)]$, such that $\dot{x}(\psi_n)(t_n) \to 0$ as $n \to \infty$. The slope $s_n^{-1}(t)$ of $Pz_t(\psi_n)$ is given by

$$s_n(t) = \frac{y(\psi_n)(t) - F(x(\psi_n)(t))}{-g(x(\psi_n)(t-r))} > 0$$

and is clearly bounded for all n. Since $t_3(\psi_n)$ is bounded, we may assume $t_n \to t_0$, $x(\psi_n)(t_n) \to x_0 > 0$, $x(\psi_n)(t_n - r) \to x_1 > 0$, $y(\psi_n)(t_n) \to y_0$ as $n \to \infty$. Therefore, $s_n(t_n) \to s_0 \geq 0$ as $n \to \infty$. The constant s_0 cannot be zero since this would imply $\dot{x}(\psi_n)(t) = 0$ for some $n \geq n_0$ sufficiently large and $t \in [t_2(\psi_n), t_3(\psi_n)]$. Thus, $s_0 > 0$. But this contradicts the fact that $x(\psi_n)(t_n) - F(x(\psi_n)(t_n)) \to \infty$ as $n \to \infty$ and proves the lemma.

<u>Lemma 31.3</u>. If $\psi \in K_0$ is an eigenfunction of A, $A\psi = \mu\psi$, then there is an $M > 0$ such that $|\psi| = M$ implies $0 < \mu < 1$.

<u>Proof</u>. We use the notation of the proof of Lemma 31.1. Suppose $\psi \in K_0$, $A\psi = \mu\psi$, and $P\psi$ is on or below Γ. Then $PA\psi$ is above Γ. Since $PA\psi = \mu P\psi$, this implies $0 < \mu < 1$. Therefore the lemma is true in this case.

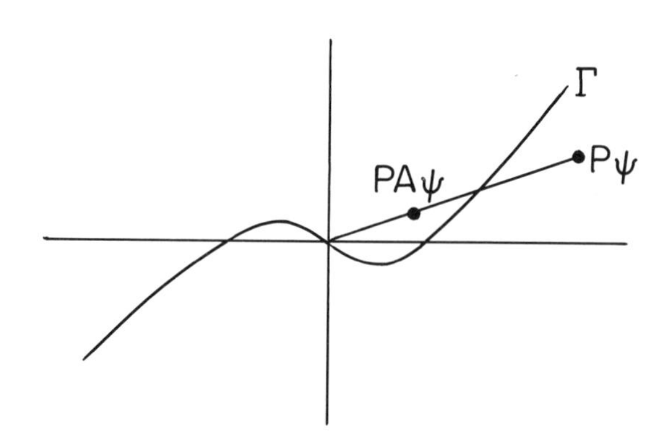

Suppose now that $P\psi$ lies above Γ and t_1, t_2, t_3, t_4 are as in the proof of Lemma 31.1 and m_0, δ are the constants given in Lemma 31.2. If $\psi = (\varphi, a) \in K_0$, $|\psi| = M \geq m_0$, $A\psi = \mu\psi$, then $-y(\psi)(t_4) = \mu a$. If we omit the argument and use (31.2), then

$$t_3 - t_2 = \int_{t_2}^{t_3} dt = \int_{x(t_2)}^{0} \frac{dx}{y-F(x)} < \frac{x(t_2)}{\delta}$$

$$\mu a = |y(t_4)| = \int_{t_2}^{t_4} g(x(t-r))dt < g(x(t_2))(t_3-t_2+r)$$

$$< x(t_2)[\frac{g(x(t_2))}{\delta} + r] < x(t_2)g(x(t_2))b, \ b = \frac{1}{\delta} + \frac{r}{g(\beta)}$$

since $x(t_2) > \beta$, $g(x(t_2)) > g(\beta)$. Since $F(x)$ is monotone increasing and approaches ∞ as $x \to \infty$, it follows that there is a unique y_2 such that $F^{-1}(y_2) = x(t_2)$. Also, $y_2 < a$. From hypothesis (d), there is a $\gamma > 0$ such that $F^{-1}(a)g(F^{-1}(a))b < a$ for $a > \gamma$. Thus,

$$\mu a = |y(t_4)| < x(t_2)g(x(t_2))b = F^{-1}(y_2)g(F^{-1}(y_2)b < F^{-1}(a)g(F^{-1}(a))b < a.$$

This shows $\mu < 1$ and proves the lemma.

Lemma 31.4. For any $r > 0$, there is a $k_0(r) > 0$ such that for any $k \geq -k_0(r)$ there is at least one pair of roots of

$$\lambda^2 - k\lambda + e^{-\lambda r} = 0 \tag{31.3}$$

lying in the region Re $\lambda > 0$, $-\pi/r < \text{Im } \lambda < \pi/r$.

Proof. If $h(\lambda) = \lambda^2 - k\lambda + e^{-\lambda r}$, $\lambda = \gamma + i\sigma$, and $\theta = \arg h$, then

$$\tan \theta = \frac{\text{Im } h}{\text{Re } h} = \frac{(2\gamma - k)\sigma - e^{-\gamma r}\sin \sigma r}{\gamma^2 - \sigma^2 - k\gamma + e^{-\gamma r}\cos \sigma r}$$

and $\tan \theta$ is an odd function of σ. Consider the contour indicated below

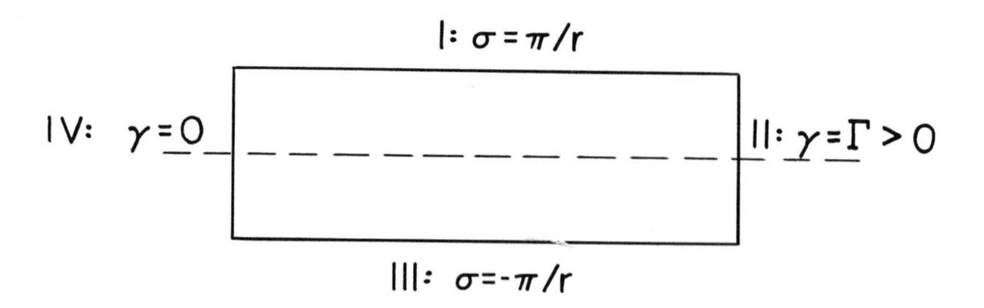

where Γ will be chosen large in a moment.

On I, we have

$$\tan\theta = \frac{(2\gamma - k)\pi/r}{\gamma^2 - k\gamma - \pi^2/r^2 - e^{-\gamma r}} .$$

The denominator is < 0 for $\gamma = 0$ and has at most one zero and the numerator has at most one zero with no zero if $k < 0$. Furthermore, the zero of the denominator is greater than $k/2$ if $k > 0$. Therefore, on I, for Γ sufficiently large, $\tan\theta$ looks like

I:
− ∞ +
k< 0
+ 0 − ∞ +
k >0

On II, we have

$$\tan\theta = \frac{(2\Gamma - k)\sigma - e^{-\Gamma r}\sin\sigma r}{\Gamma^2 - k\Gamma - \sigma^2 + e^{-\Gamma r}\cos\sigma r}$$

and for Γ large the denominator is positive for any fixed k and the numerator will not have any zero in $0 < |\sigma| < \pi/r$ except $\sigma = 0$. Therefore, on II, $\tan\theta$

looks like

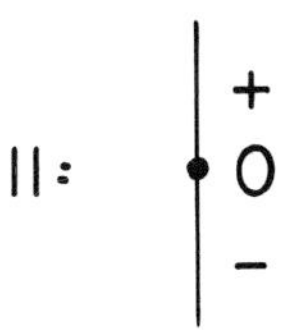

On IV, we have

$$\tan\theta = \frac{-k\sigma - \sin\sigma r}{-\sigma^2 + \cos\sigma r}.$$

The denominator vanishes at two points for $\sigma \in [-\pi/r, \pi/r]$ and the numerator has exactly one zero at $\sigma = 0$ if either $k > 0$ or $k \leq -r$. If $-r < k < 0$, then there are three zeros. Therefore, $\tan\theta$ on IV looks like

IV: $k>0$ — $+ \; \infty \; - \; 0 \; + \; \infty \; -$; $k \leq -r$ — $- \; \infty \; + \; 0 \; - \; \infty \; +$; $-r < k < 0$ — $- \; \infty \; + \; 0 \; - \; 0 \; + \; 0 \; - \; \infty \; +$ or $- \; 0 \; + \; \infty \; - \; 0 \; + \; \infty \; - \; 0 \; +$

If we use the fact that $\tan\theta$ is odd in σ, then we can count the change $\Delta \arg h = \Delta\theta$ in the argument of the function h as we traverse the contour under investigation. For $k > 0$, $\Delta\theta = 4\pi$, and there

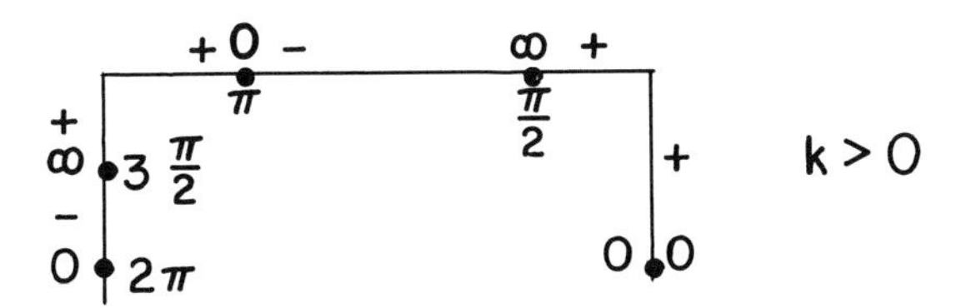

are two roots enclosed by the contour. If $k \leq -r$, then

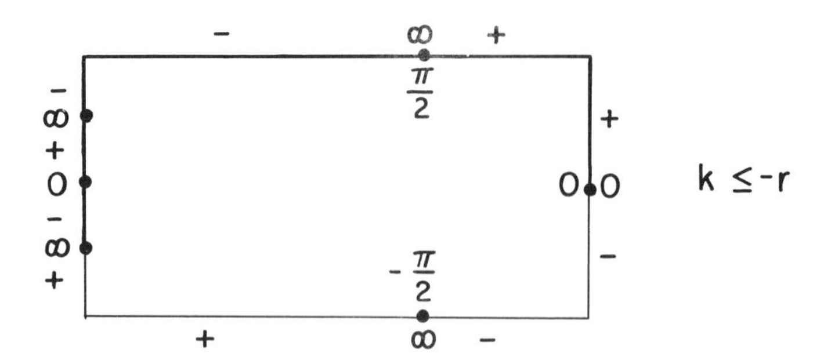

there are no roots since $\Delta\theta = 0$. If $-r < k < 0$, then either there are no roots or there are two roots.

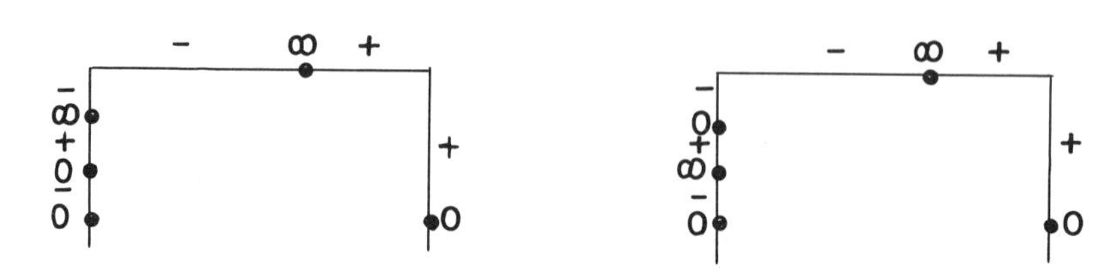

The case where two roots occur is the zero $\sigma_0(r)$ of $\sigma^2 = \cos \sigma r$, $0 < \sigma_0(r) < \pi/r$ is less than the zero of $-k = (\sin \sigma r)/r$. If we let

$$k_0(r) = [\sin \sigma_0(r) r]/r$$

then the lemma is proved for this $k_0(r)$.

Note that $k_0(r) \to 0$ as $r \to 0$ and in fact

$$k_0(r) = r - \frac{1}{6} r^2 + \cdots$$

for r small.

Let π_U be the same operator as in the previous examples in Section 29 and 30.

<u>Lemma 31.5</u>. If $k_0(r)$ is as in Lemma 31.4, then $\inf_{\varphi \in \partial B(1) \cap K_0} |\pi_U \varphi| > 0$ if $f(0) < k_0(r)$.

<u>Proof</u>. For $k = -f(0)$ the equation adjoint to the linear equation

$$\dot{x}(t) = y(t) + kx(t)$$

$$\dot{y}(t) = -x(t-r)$$

is

$$\dot{w}(t) = -w(t)A - w(t+r)B$$

$$A = \begin{bmatrix} k & 1 \\ 0 & 0 \end{bmatrix}, \quad B = \begin{bmatrix} 0 & 0 \\ -1 & 0 \end{bmatrix}$$

and the bilinear form is

$$(\zeta, \psi) = \zeta(0)\psi(0) + \int_{-r}^{0} \zeta(\xi + r) \begin{bmatrix} 0 \\ -\varphi(\xi) \end{bmatrix} d\xi$$

where we have let $\psi = (\varphi, a)$, $\varphi \in C$, $a \in R$. As in the preceding application, to prove the lemma it is sufficient to consider only the projections onto the eigenvalues with positive real parts given by Lemma 31.4.

If $\zeta(t) = e^{-\lambda t} b$ is a solution of the adjoint equation, then one can choose

$$b = \left(\frac{1}{k-\lambda}, -1\right).$$

Let us consider first the case where $\lambda = \gamma + i\sigma$, $0 < \sigma < \pi/r$. If $b = \alpha - i\beta$ where α, β are real two vectors, then the real and imaginary parts of (ζ, ψ) are given by

$$\mathrm{Re}(\zeta,\psi) = \alpha\psi(0) + \int_{-r}^{0} e^{-\gamma(\xi-r)}\cos\sigma(\xi+r)\varphi(\xi)d\xi$$

$$-\mathrm{Im}(\zeta,\psi) = \beta\psi(0) + \int_{-r}^{0} e^{-\gamma(\xi-r)}\sin\sigma(\xi+r)\varphi(\xi)d\xi$$

$$\beta\psi(0) = \frac{\sigma}{|\lambda-k|}\varphi(0) \geq 0.$$

If there is a sequence $\psi_n = (\varphi_n, a_n) \in \partial B(1) \cap K_0$ such that $\pi_U\psi_n \to 0$ as $n \to \infty$, then $\mathrm{Im}(\zeta,\psi_n) \to 0$ as $n \to \infty$. But this implies $\varphi_n(\xi) \to 0$ as $n \to \infty$. Therefore, $\mathrm{Re}(\zeta,\psi_n) - a_n \to 0$ as $n \to \infty$. Since $\mathrm{Re}(\zeta,\psi_n) \to 0$, this yields $a_n \to 0$ as $n \to \infty$. Therefore $\psi_n \to 0$ as $n \to \infty$, which is a contradiction.

Suppose now that $\lambda_1 \geq \lambda_2$ are real positive roots of (31.3) and $\zeta_1(t) = e^{-\lambda_1 t}b$, $\zeta_2(t) = e^{-\lambda_2 t}c$ are solutions of the adjoint equation with

$$b = \left(\frac{1}{k-\lambda_1}, -1\right), \qquad c = \left(\frac{1}{k-\lambda_2}, -1\right).$$

For this choice of b, c,

$$(\zeta_1,\psi) = b\psi(0) + \int_{-r}^{0} e^{-\lambda_1(\xi+r)}\varphi(\xi)d\xi = \frac{\varphi(0)}{k-\lambda_1} - a + \int_{-r}^{0} e^{-\lambda_1(\xi+r)}\varphi(\xi)d\xi$$

$$(\zeta_2,\psi) = c\psi(0) + \int_{-r}^{0} e^{-\lambda_2(\xi+r)}\varphi(\xi)d\xi = \frac{\varphi(0)}{k-\lambda_2} - a + \int_{-r}^{0} e^{-\lambda_2(\xi+r)}\varphi(\xi)d\xi.$$

Suppose there is a sequence $\psi_n = (\varphi_n, a_n) \in \partial B(1) \cap K_0$ such that $(\zeta_1,\psi_n) \to 0$,

$(\zeta_2, \psi_n) \to 0$ as $n \to \infty$. We can choose this sequence so that $\varphi_n(0) \to \varphi_0$, $a_n \to a_0$ as $n \to \infty$. Then

$$(31.4) \qquad \varphi_0\left(\frac{1}{k-\lambda_1} - \frac{1}{k-\lambda_2}\right) + \int_{-r}^{0} [e^{-\lambda_1(\xi+r)} - e^{-\lambda_2(\xi+r)}]\varphi_n(\xi)d\xi \to 0$$

as $n \to \infty$.

Since $\lambda_1 \geq \lambda_2$, we have

$$0 \leq \int_{-r}^{0} [e^{-\lambda_2(\xi+r)} - e^{-\lambda_1(\xi+r)}]\varphi_n(\xi)d\xi = \int_{0}^{r} [e^{-\lambda_2 u} - e^{-\lambda_1 u}]\varphi_n(u-r)du$$

$$\leq \varphi_n(0)[-\frac{1}{\lambda_2} e^{-\lambda_2 r} + \frac{1}{\lambda_2} + \frac{1}{\lambda_1} e^{-\lambda_1 r} - \frac{1}{\lambda_1}]$$

$$= \varphi_n(0)[\lambda_2 - k + \frac{1}{\lambda_2} - (\lambda_1 - k) - \frac{1}{\lambda_1}] = \varphi_n(0)[\lambda_2 - \lambda_1 + \frac{\lambda_1-\lambda_2}{\lambda_1\lambda_2}]$$

$$= \varphi_n(0)\,\frac{(\lambda_1-\lambda_2)}{\lambda_1\lambda_2}(1 - \lambda_1\lambda_2).$$

On the other hand,

$$\frac{\varphi_0(\lambda_1-\lambda_2)}{(k-\lambda_1)(k-\lambda_2)}[1 - \frac{(k-\lambda_1)}{\lambda_2}\frac{(k-\lambda_2)}{\lambda_1} + (k-\lambda_1)(k-\lambda_2)] > \varphi_0(\lambda_1-\lambda_2)$$

since $k < \lambda_1 + \lambda_2$. The latter inequality is valid from the accompanying diagram and the fact that $\beta > \alpha$. Since (31.4) must hold, this implies $\varphi_0 = 0$, and as a consequence $\varphi_n(\xi) \to 0$ as $n \to \infty$. Returning to the expression for (ζ_1, ψ_n), we have $a_n \to 0$ and thus $\psi_n \to \infty$. This is a contradiction. The case where $\lambda_1 = \lambda_2$ and the independent solutions of the adjoint equation are $\zeta_1(t) = e^{\lambda_1 t} b$, $\zeta_2(t) = e^{\lambda_1 t}(bt+c)$ remains to be considered. Following the same procedure as above, the reader can easily supply the details. This will complete the proof of the lemma.

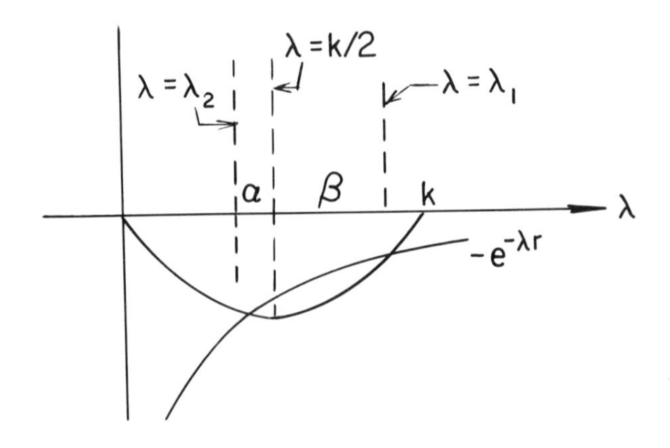

Using the above lemmas and Theorem 28.1, we have

Theorem 31.1. If F, g satisfy conditions (a) - (d) and $f(0) < k_0(r)$ where $k_0(r) > 0$ is given in Lemma 31.4, then Equation (31.1) has a nonconstant periodic solution.

The proof of the above theorem is modeled after the one of Grafton for the case $f(x) = k(x^2-1)$, $k > 0$; that is, the van der Pol equation with a retardation. It is interesting to look at the latter equation in more detail. If $x = u/\sqrt{k}$

$$\ddot{u}(t) + (u^2(t) - k)\dot{u}(t) + u(t-r) = 0.$$

This equation satisfies the conditions of Theorem 31.1 for every $k > 0$. Therefore, there is a periodic solution $u^*(k)$ with $|u^*(k)| \geq c > 0$ for $0 \leq k \leq 1$. Thus, the solution $x^*(k) = u^*(k)/\sqrt{k}$ of

$$\ddot{x}(t) + k(x^2(t)-1)\dot{x}(t) + x(t-r) = 0$$

approaches ∞ as $k \to 0$.

The conditions of Theorem 31.1 are also satisfied by $f(x) = ax^2 + b$, $a > 0$, $b < k_0(r)$, $g(x) = x$, and, in particular, for $f(x) = x^2$, $g(x) = x$.

32. THE "ADJOINT" EQUATION FOR GENERAL LINEAR SYSTEMS

In this section, we consider the general linear system

$$\dot{x}(t) = \int_{-r}^{0} [d_\theta \eta(t,\theta)]x(t+\theta) \overset{\text{def}}{=} L(t,x_t) \tag{32.1}$$

where $\eta(\cdot,\cdot)$ is an $n \times n$ matrix valued function which is measurable in $(t,\theta) \in R \times R$ and normalized so that

$$\eta(t,\theta) = 0 \quad \text{for} \quad \theta \geq 0, \quad \eta(t,\theta) = \eta(t,-r) \quad \text{for} \quad \theta \leq -r.$$

Furthermore, we assume that $\eta(t,\theta)$ is of bounded variation in θ for each t with $\mathrm{Var}_{[-r,0]}\eta(t,\cdot) \leq m(t)$ where $m(\cdot)$ is locally integrable on R, and $\eta(t,\theta)$ is continuous from the left in θ on $(-r,0)$.

Along with the homogeneous equation (32.1), we consider the nonhomogeneous equation

$$\dot{x}(t) = L(t,x_t) + h(t) \tag{32.2}$$

where h is a locally integrable function from R into E^n.

The purpose of this section is to generalize the results of Section 17 to system (32.2). More specifically, Theorem 16.3 implies that the solution $x(\sigma,\varphi)$ of (32.2) through (σ,φ) can be written as

$$x(\sigma,\varphi)(t) = y(\sigma,\varphi)(t) + \int_{\sigma}^{t} U(t,s)h(s)ds, \quad t \geq \sigma, \tag{32.3}$$

where $y(\sigma,\varphi)$ is the solution of the homogeneous equation (32.1) through (σ,φ) and $U(t,s)$ satisfies (16.11). Our objective is to obtain a representation of $y(\sigma,\varphi)(t)$ through $U(t,s)$ as well as a certain matrix solution of the "adjoint" equation

$$(32.4) \qquad z(s) + \int_s^\infty z(\alpha)\eta(\alpha, s-\alpha)d\alpha = \text{constant},$$

where z is in E^{n*}.

This definition of the "adjoint" equation may look as if it is different than the one given for the special case in Section 17. To see that this is not the case, we consider only a special case but the general case can be treated in the same manner. For the equation

$$\dot{x}(t) = A(t)x(t) + B(t)x(t-r) \overset{\text{def}}{=} \int_{-r}^{0} [d_\theta \eta(t,\theta)]x(t+\theta),$$

the "adjoint" of Section 17 is

$$\dot{z}(s) = -z(s)A(s) - z(s+r)B(s+r).$$

The normalized function η for this equation is

$$\eta(t,\theta) = \begin{cases} -A(t) - B(t), & \theta \leq -r \\ -A(t) & -r < \theta < 0 \\ 0 & \theta = 0. \end{cases}$$

Therefore, the "adjoint" (32.4) is

$$z(s) + \int_s^{s+r} [-z(\alpha)A(\alpha)]d\alpha + \int_{s+r}^{\infty} [-z(\alpha)][A(\alpha)+B(\alpha)]d\alpha = \text{constant}$$

which reduces to the differential equation above for z.

Let B_0 denote the Banach space of functions $\psi\colon [-r,0] \to E^{n*}$ of bounded variation on $[-r,0]$, continuous from the left on $(-r,0)$ and vanishing at zero with norm $\text{Var}_{[-r,0]}\psi$. A straightforward application of the contraction mapping principle yields

<u>Theorem 32.1</u>. For any $t \in R$, $\psi \in B_0$, there exists a unique $z\colon R \to E^{n*}$ which

vanishes on $[t,\infty)$, satisfies (32.4) on $(-\infty,t-r]$ and such that $z_t = \psi$. For any $s \leq t$, let z_s^0 be defined by

$$z_s^0(\theta) = z(s+\theta), \quad -r \leq \theta < 0, \quad z_s^0(0) = 0.$$

Then this solution may be written for each $s \leq t$ as

$$z_s^0 = \tilde{T}(s,t)z_t^0 = \tilde{T}(s,t)\psi$$

where $\tilde{T}(s,t): B_0 \to B_0$ is bounded and linear.

The representation theorem for the solutions of (32.2) is contained in

<u>Theorem 32.2</u>. If x satisfies (32.2) on $[\sigma,\infty)$, then for any $t \geq \sigma$,

$$x(t) = Y(\sigma,t)x(\sigma) + \int_{\sigma-r}^{\sigma^-} d_\beta\{\int_\sigma^t Y(\alpha,t)\eta(\alpha,\beta-\alpha)d\alpha\}x(\beta) + \int_\sigma^t Y(\alpha,t)h(\alpha)d\alpha \tag{32.5}$$

where the $n \times n$ matrix valued function Y is defined by

$$\begin{aligned} Y(\sigma,t) &= 0 \quad \text{for} \quad \sigma > t \\ Y(\sigma,t) &= I - \int_\sigma^t Y(\alpha,t)\eta(\alpha,\sigma-\alpha)d\alpha \quad \text{for} \quad \sigma \leq t. \end{aligned} \tag{32.6}$$

Furthermore, $Y(\sigma,t)$ is absolutely continuous in t (except at $t = \sigma$), locally of bounded variation in σ and $Y(\sigma,t) = U(t,\sigma)$ a.e. in σ, where $U(t,\sigma)$ is defined in (16.11).

<u>Proof</u>. From Theorem 32.1, there is a solution $Y(\sigma,t)$ of (32.6), locally of bounded variation in σ. Let $W(\sigma,t) = 0$ for $\sigma > t$,

$$W(\sigma,t) = -\eta(t,\sigma-t) - \int_\sigma^t W(\alpha,t)\eta(\alpha,\sigma-\alpha)d\alpha.$$

Then one easily shows that

$$|W(\sigma,t)| \leq \mathrm{Var}_{[\sigma,t]} W(\cdot,t) \leq m(t)\exp[\int_\sigma^t m(\alpha)d\alpha].$$

Furthermore, for $t \geq \sigma$,

$$\begin{aligned} I + \int_\sigma^t W(\sigma,\tau)d\tau &= I + \int_\sigma^t [-\eta(\tau,\sigma-\tau) - \int_\sigma^\tau W(\alpha,\tau)\eta(\alpha,\sigma-\alpha)d\alpha]d\tau \\ &= I - \int_\sigma^t \eta(\alpha,\sigma-\alpha)d\alpha - \int_\sigma^t (\int_\alpha^t W(\alpha,\tau)d\tau)\eta(\alpha,\sigma-\alpha)d\alpha \\ &= I - \int_\sigma^t [I + \int_\alpha^t W(\alpha,\tau)d\tau]\eta(\alpha,\sigma-\alpha)d\alpha \end{aligned}$$

and $I + \int_\sigma^t W(\sigma,\tau)d\tau$ satisfies the same equation as $Y(\sigma,t)$. Uniqueness of the solution implies $Y(\sigma,t) = I + \int_\sigma^t W(\sigma,\tau)d\tau$ for $t \geq \sigma$ is absolutely continuous in t and of bounded variation in σ.

Suppose x is the solution of (32.2) through (σ,φ). Since $x(t)$ is continuous in t, the following integration by parts is valid,

$$\int_\sigma^{t^+} d_\alpha Y(\alpha,t)x(\alpha) + \int_\sigma^{t^+} Y(\alpha,t)d_\alpha x(\alpha) = -Y(\sigma,t)x(\sigma).$$

Using the fact that x satisfies (32.2) and is absolutely continuous for $t \geq \sigma$, the second integral is the same as the integral from σ to t. The first integral is equal to $-x(t) + \int_\sigma^t d_\alpha Y(\alpha,t)x(\alpha)$. Using all of this information, multiplying (32.2) by $Y(\alpha,t)$ and integrating from σ to t, we have the identity

$$x(t) - Y(\sigma,t)x(\sigma) - \int_\sigma^t Y(\alpha,t)h(\alpha)d\alpha - \int_\sigma^t d_\alpha Y(\alpha,t)x(\alpha) = \int_\sigma^t Y(\alpha,t)L(\alpha,x_\alpha)d\alpha. \tag{32.7}$$

Using the fact that $\eta(\alpha,\theta) = \eta(\alpha,-r)$ for $\theta \leq -r$, $\eta(\alpha,\theta) = 0$ for $\theta \geq 0$, we have

$$\int_{\sigma}^{t} Y(\alpha,t)L(\alpha,x_{\alpha})d\alpha = \int_{\sigma}^{t} \{Y(\alpha,t)\int_{\sigma-r}^{t} [d_{\beta}\eta(\alpha,\beta-\alpha)]x(\beta)\}d\alpha$$

$$= \int_{\sigma-r}^{t} d_{\beta}\{\int_{\sigma}^{t} Y(\alpha,t)\eta(\alpha,\beta-\alpha)d\alpha\}x(\beta),$$

where we have used an unsymmetric Fubini theorem of Cameron and Martin. Since $\eta(\alpha,\beta-\alpha) = 0$ for $\beta \geq \alpha$, one can write the expression above in the following manner

$$\int_{\sigma}^{t} Y(\alpha,t)L(\alpha,x_{\alpha})d\alpha = \int_{\sigma-r}^{\sigma^-} d_{\beta}\{\int_{\sigma}^{t}\}x(\beta) + \int_{\sigma}^{t} d_{\beta}\{\int_{\beta}^{t}\}x(\beta).$$

Now using the fact that Y satisfies (32.6), we have

$$\int_{\sigma}^{t} Y(\alpha,t)L(\alpha,x_{\alpha})d\alpha = \int_{\sigma-r}^{\sigma^-} d_{\beta}\{\int_{\sigma}^{t}\}x(\beta) + \int_{\sigma}^{t} d_{\beta}[I-Y(\beta,t)]x(\beta)$$

$$= \int_{\sigma-r}^{\sigma^-} d_{\beta}\{\int_{\sigma}^{t}\}x(\beta) - \int_{\sigma}^{t} d_{\beta}Y(\beta,t)x(\beta).$$

Using this expression in (32.7), one arrives at (32.5).

For the initial value $(\sigma,0)$, we have

$$x(t) = \int_{\sigma}^{t} Y(\alpha,t)h(\alpha)d\alpha$$

for every h which is locally integrable. But Theorem 16.3 implies that $x(t) = \int_{\sigma}^{t} U(t,\alpha)h(\alpha)d\alpha$ where U is defined in (16.11). Therefore, $Y(\alpha,t) = U(t,\alpha)$ a.e. in α. This proves the theorem.

33. THE TRUE ADJOINT OF A LINEAR SYSTEM

In this section, we consider the linear systems (32.1), (32.2) under the same hypotheses as in Section 32. We will identify B_0 with the conjugate space of C using the pairing

$$(33.1) \qquad \langle \psi, \varphi \rangle = \int_{-r}^{0} [d\psi(\theta)]\varphi(\theta) \qquad \text{for } \psi \in B_0,\ \varphi \in C.$$

If we designate the solution of (32.2) through (σ,φ) by $x(\sigma,\varphi,h)$, then

$$(33.2) \qquad x_t(\sigma,\varphi,h) = T(t,\sigma)\varphi + K(t,\sigma)h$$

where $T(t,\sigma)\colon C \to C$, $K(t,\sigma)\colon \mathcal{L}_1([\sigma,t],E^n) \to C$, $t \geq \sigma$, are continuous linear operators with $T(\sigma,\sigma) = I$ and $K(\sigma,\sigma) = 0$. The adjoints of these operators are defined by

$$T^*(\sigma,t)\colon B_0 \to B_0,\ K^*(\sigma,t)\colon B_0 \to \mathcal{L}_\infty([\sigma,t],E^{n*})$$

$$(33.3) \qquad \langle T^*(\sigma,t)\psi,\varphi \rangle = \langle \psi, T(t,\sigma)\varphi \rangle$$

$$(33.4) \qquad \int_{\sigma}^{t} (K^*(\sigma,t)\psi)(s)h(s)ds = \langle \psi, K(t,\sigma)h \rangle$$

whenever $\varphi \in C$, $\psi \in B_0$, $h \in \mathcal{L}_1([\sigma,t],E^n)$ and $t \geq \sigma$.

The main theorem of this section is

Theorem 33.1. For any $t \geq \sigma$,

$$(33.5) \qquad T^*(\sigma,t) = (I+\Omega(\sigma))\tilde{T}(\sigma,t)(I+\Omega(t))^{-1}$$

where $\tilde{T}(\sigma,t)$ is given in Theorem 32.1 and $\Omega(\sigma)$ is the quasi-nilpotent operator on B_0 defined by

$$(33.6)\qquad (\Omega(\sigma)\psi)(\theta) = \int_\theta^0 \psi(\alpha)\eta(\sigma+\alpha,\theta-\alpha)d\alpha, \qquad -r \le \theta \le 0$$

for any $\sigma \in R$, $\psi \in B_0$.

Proof: For any $\varphi \in C$, $\psi \in B_0$, let $x = x(\sigma,\varphi,0)$ so that $x_\sigma = \varphi$, $x_t = T(t,\sigma)\varphi$ for $t \ge \sigma$. Extend ψ by setting $\psi(\xi) = \psi(-r)$ for $\xi \le -r$, $\psi(\xi) = 0$ for $\xi \ge 0$ and extend x in any continuous manner to $(-\infty,\infty)$. From Theorem 32.2,

$$\langle T^*(\sigma,t)\psi,\varphi\rangle = \langle \psi, x_t\rangle = \int_{-\infty}^0 [d\psi(\theta)]x(t+\theta)$$

$$= \int_{\sigma-t-r}^{(\sigma-t)^-} [d\psi(\theta)]\varphi(\theta+t+\sigma) + \int_{\sigma-t}^0 [d\psi(\theta)]x(t+\theta)$$

where $x(t+\theta)$ in the second integral is given in (32.5). Therefore,

$$(33.7)\qquad \langle T^*(\sigma,t)\psi,\varphi\rangle = \int_{-r}^{0^-} [d_\xi\psi(\xi+\sigma-t)]\varphi(\xi) + \int_{\sigma-t}^0 [d\psi(\theta)]Y(\sigma,t+\theta)\varphi(0)$$

$$+ \int_{\sigma-t}^0 [d\psi(\theta)] \int_{-r}^0 d_\xi\{\int_\sigma^t Y(\alpha,t+\theta)\eta(\alpha,\sigma+\xi-\alpha)d\alpha\}\varphi(\xi).$$

Interchanging the order of integration in the last integral, we obtain

$$\int_{-r}^0 d_\xi\{\int_{\sigma-t}^0 [d\psi(\theta)]\int_\sigma^t Y(\alpha,t+\theta)\eta(\alpha,\sigma+\xi-\alpha)d\alpha\}\varphi(\xi)$$

$$= \int_{-r}^0 d_\xi\{\int_\sigma^t [\int_{\sigma-t}^0 (d\psi(\theta))Y(\alpha,t+\theta)]\eta(\alpha,\sigma+\xi-\alpha)d\alpha\}\varphi(\xi)$$

$$= \int_{-r}^0 d_\xi\{\int_\sigma^t [\int_{\alpha-t}^0 (d\psi(\theta))Y(\alpha,t+\theta)]\eta(\alpha,\sigma+\xi-\alpha)d\alpha\}\varphi(\xi)$$

since $Y(\sigma,t) = 0$ for $\sigma > t$. If we define a function $y(\alpha,t) = 0$ for $\alpha > t$ and

$$(33.8)\qquad y(\alpha,t) = -\int_{\alpha-t}^0 [d\psi(\theta)]Y(\alpha,t+\theta), \quad \alpha \le t,$$

then using the above computations and (33.7), we have

$$(33.9) \qquad \langle T^*(\sigma,t)\psi,\varphi\rangle = \int_{-r}^{0^-} [d_\xi \psi(\xi+\sigma-t)]\varphi(\xi) - y(\sigma,t)\varphi(0)$$

$$- \int_{-r}^{0} d_\xi \{\int_\sigma^t y(\alpha,t)\eta(\alpha,\sigma+\xi-\alpha)d\alpha\}\varphi(\xi).$$

Since relation (33.9) must hold for any $\varphi \in C$, it follows from (33.9) and the definition of $\langle\cdot,\cdot\rangle$ that $[T^*(\sigma,t)\psi](\theta)$ has a jump at $\theta = 0$ and, in particular,

$$(33.10) \qquad [T^*(\sigma,t)\psi][0^-] = y(\sigma,t).$$

From (33.8) and (32.6), we see that, for $t \geq \sigma$,

$$\begin{aligned}
y(\sigma,t) &= -\int_{\sigma-t}^{0} [d\psi(\theta)]Y(\sigma,t+\theta) \\
&= -\int_{\sigma-t}^{0} d\psi(\theta) + \int_{\sigma-t}^{0} [d\psi(\theta)]\int_\sigma^{t+\theta} Y(\alpha,t+\theta)\eta(\alpha,\sigma-\alpha)d\alpha \\
&= \psi(\sigma-t) + \int_{\sigma-t}^{0} [d\psi(\theta)]\int_\sigma^{t} Y(\alpha,t+\theta)\eta(\alpha,\sigma-\alpha)d\alpha \\
&= \psi(\sigma-t) + \int_\sigma^t \{\int_{\sigma-t}^{0} [d\psi(\theta)]Y(\alpha,t+\theta)\}\eta(\alpha,\sigma-\alpha)d\alpha \\
&= \psi(\sigma-t) + \int_\sigma^t \{\int_{\alpha-t}^{0} [d\psi(\theta)]Y(\alpha,t+\theta)\}\eta(\alpha,\sigma-\alpha)d\alpha \\
&= \psi(\sigma-t) - \int_\sigma^t y(\alpha,t)\eta(\alpha,\sigma-\alpha)d\alpha.
\end{aligned}$$

Therefore,

$$(33.11) \qquad y(\sigma,t) + \int_\sigma^t y(\alpha,t)\eta(\alpha,\sigma-\alpha)d\alpha = \psi(\sigma-t), \quad \sigma \leq t.$$

Since $\psi(\sigma-t) = \psi(-r)$ for $\sigma \leq t - r$, it follows that

$$y(\sigma,t) + \int_\sigma^t y(\alpha,t)\eta(\alpha,\sigma-\alpha)d\alpha = \psi(-r), \qquad \sigma \leq t - r,$$

and $y(\sigma,t)$ as a function of σ is a solution of the "adjoint" equation (32.4). From Theorem 32.1, we can, therefore, write

$$y^0_\sigma(\cdot,t) = \tilde{T}(\sigma,t)y^0_t(\cdot,t).$$

The initial function $y^0_t(\cdot,t)$ is given from (33.11) by

$$y(t+\theta,t) + \int_{t+\theta}^{t} y(\alpha,t+\theta)\eta(\alpha,t+\theta-\alpha)d\alpha = \psi(\theta), \quad -r \leq \theta \leq 0$$

or

$$y(t+\theta,t) + \int_{\theta}^{0} y(t+\alpha,t+\theta)\eta(t+\alpha,\theta-\alpha)d\alpha = \psi(\theta), \quad -r \leq \theta \leq 0.$$

Comparing with (33.6), we have $[I+\Omega(t)]y^0_t(\cdot,t) = \psi$ and

$$y^0_\sigma(\cdot,t) = \tilde{T}(\sigma,t)(I+\Omega(t))^{-1}\psi, \quad \sigma \leq t. \tag{33.12}$$

Returning to (33.9), using (33.11) and the normalization condition (33.10), for $-r \leq \theta < 0$,

$$\begin{aligned}
[T^*(\sigma,t)\psi](\xi) &= \psi(\xi+\sigma-t) - \int_{\sigma}^{t} y(\alpha,t)\eta(\alpha,\sigma+\xi-\alpha)d\alpha \\
&= y(\sigma+\xi,t) + \int_{\sigma+\xi}^{\sigma} y(\alpha,t)\eta(\alpha,\sigma+\xi-\alpha)d\alpha \\
&= y(\sigma+\xi,t) + \int_{\xi}^{0} y(\sigma+\beta,t)\eta(\sigma+\beta,\xi-\beta)d\beta \\
&= [(I+\Omega(\sigma))y^0_\sigma(\cdot,t)](\xi)
\end{aligned}$$

from (33.6). Using (33.12) in this expression, the proof of the theorem is complete.

<u>Corollary 33.1.</u> The adjoint $K^*(\sigma,t)$ of $K(t,\sigma)$ defined by (33.4) may be written as

$$[K^*(\sigma,t)\psi](\xi) = -[T^*(\sigma,t)\psi](0^-) = -\int_{-r}^{0} [d\psi(\theta)]Y(\xi,t+\theta)$$

for almost every ξ in $[\sigma,t]$ and any $\psi \in B_0$.

<u>Proof</u>: If $h \in \mathcal{L}_1([\sigma,t],E^n)$ and $\psi \in B_0$, then, by Theorem 33.2,

$$\begin{aligned}
\int_{\sigma}^{t} (K^*(\sigma,t)\psi)(\alpha)h(\alpha)d\alpha &= \langle \psi, K(t,\sigma)h\rangle \\
&= \int_{-r}^{0} [d\psi(\theta)](K(t,\sigma)h)(\theta) \\
&= \int_{-r}^{0} [d\psi(\theta)]\int_{\sigma}^{t+\theta} Y(\alpha,t+\theta)h(\alpha)d\alpha \\
&= \int_{-r}^{0} [d\psi(\theta)]\int_{\sigma}^{t} Y(\alpha,t+\theta)h(\alpha)d\alpha \\
&= \int_{\sigma}^{t} \{\int_{-r}^{0} [d\psi(\theta)]Y(\alpha,t+\theta)\}h(\alpha)d\alpha \\
&= -\int_{\sigma}^{t} y(\alpha,t)h(\alpha)d\alpha \\
&= -\int_{\sigma}^{t} [T^*(\alpha,t)\psi](0^-)h(\alpha)d\alpha
\end{aligned}$$

from (33.8) and (33.10). Since this relation must hold for every $h \in \mathcal{L}_1([\sigma,t],E^n)$, we obtain the corollary.

34. BOUNDARY VALUE PROBLEMS

In this section, we discuss two point boundary value problems for the nonhomogeneous equation (32.2) and obtain results of the "Fredholm alternative" type. With these results, applications to weakly nonlinear problems can be obtained in the standard manner.

Suppose V is a Banach space, $\sigma < \tau$ are given real numbers, $M,N: C \to V$ are linear operators with domain dense in C and $\gamma \in V$ is fixed. The problem is to find a solution x of

$$(34.1) \qquad \dot{x}(t) = L(t,x_t) + h(t)$$

subject to the boundary condition

$$(34.2) \qquad Mx_\sigma + Nx_\tau = \gamma.$$

Let V^* be the conjugate space of V and M^*,N^* the adjoint operators of M,N respectively. The fundamental result is

<u>Theorem 34.1.</u> In order that (34.1), (34.2) have a solution, it is necessary that

$$(34.3) \qquad \int_\sigma^\tau z(\alpha)h(\alpha)d\alpha = -\langle \delta, \gamma \rangle_V$$

for all $\delta \in V^*$ and solutions z of the "adjoint" problem:

$$(34.4) \qquad z(s) + \int_s^\tau z(\alpha)\eta(\alpha, s-\alpha)d\alpha = \text{constant}, \quad \sigma - r \le s \le \tau - r,$$

$$(34.5) \qquad z_\sigma^0 = -(I+\Omega(\sigma))^{-1}M^*\delta, \quad z_\tau^0 = (I+\Omega(\tau))^{-1}N^*\delta.$$

If $\mathscr{R}(M+NT(\tau,\sigma))$ is closed, then the above condition is sufficient.

Proof: Since any solution of (34.1) is given by $x_\tau = T(\tau,\sigma)x_\sigma + K(\tau,\sigma)h$, relation (34.2) is equivalent to

$$NK(\tau,\sigma)h - \gamma \in \mathcal{R}(M+NT(\tau,\sigma)).$$

Therefore, it is necessary, and under the closure hypothesis sufficient, that

$$NK(\tau,\sigma)h - \gamma \in \overline{\mathcal{R}(M+NT(\tau,\sigma))} = \{\mathfrak{N}(M^*+T^*(\sigma,\tau)N^*)\}^\perp ;$$

that is, for any $\delta \in V^*$ such that $M^*\delta + T^*(\sigma,\tau)N^*\delta = 0$, we must have $\langle \delta, NK(\tau,\sigma)h - \gamma\rangle_V = 0$, or

$$\begin{aligned} \langle \delta,\gamma\rangle_V &= \langle \delta, NK(\tau,\sigma)h\rangle_V = \langle N^*\delta, K(\tau,\sigma)h\rangle \\ &= \int_\sigma^\tau (K^*(\sigma,\tau)N^*\delta)(s)h(s)ds \\ &= -\int_\sigma^\tau (T^*(\xi,\tau)N^*\delta(0^-)h(\xi)d\xi \end{aligned} \tag{34.6}$$

from Corollary 33.1.

If z is a solution of the adjoint equation (34.4) on $[\sigma-r,\tau-r]$ satisfying $z^0_\tau = (I+\Omega(\tau))^{-1}N^*\delta$, then

$$\begin{aligned} z^0_\sigma &= \tilde{T}(\sigma,\tau)z^0_\tau = \tilde{T}(\sigma,\tau)(I+\Omega(\tau))^{-1}N^*\delta \\ &= (I+\Omega(\sigma))^{-1}T^*(\sigma,\tau)N^*\delta = -(I+\Omega(\sigma))^{-1}M^*\delta \end{aligned}$$

since $\delta \in \mathfrak{N}(M^*+T^*(\sigma,\tau)N^*)$. The existence of a $\delta \in \mathfrak{N}(M^*+T^*(\sigma,\tau)N^*)$ is equivalent to the existence of a solution of (34.4) on $[\sigma-r,\tau-r]$ which satisfies (34.5). Also, for $\sigma \le \xi < \tau$, $T^*(\xi,\tau)N^*\delta = (I+\Omega(\xi))\tilde{T}(\xi,\tau)(I+\Omega(\tau))^{-1}N^*\delta =$ $(I+\Omega(\xi))\tilde{T}(\xi,\tau)z^0_\tau$ and, thus,

$$(T^*(\xi,\tau)N^*\delta)(0^-) = (\tilde{T}(\xi,\tau)z^0_\tau)(0^-) = z(\xi).$$

Substitution in (34.6) yields (34.3) and proves the theorem.

As another boundary value problem, suppose $P,Q\colon V \to C$ are linear operators with domain dense in V, p,q are fixed elements of C. The problem is to find a $\nu \in V$ and a solution of (34.1) on $[\sigma,\tau]$ such that

(34.7) $$x_\sigma = P\nu + p, \qquad x_\tau = Q\nu + q.$$

Let P^*,Q^* be the adjoints of P,Q, respectively.

Theorem 34.2. In order for (34.1), (34.7) to have a solution, it is necessary that

(34.8) $$\int_\sigma^\tau z(\xi)h(\xi)d\xi = \langle (I+\Omega(\sigma))z_\sigma^0, p\rangle - \langle (I+\Omega(\tau))z_\tau^0, q\rangle$$

for every solution z of the adjoint problem on $[\sigma-r,\tau-r]$ satisfying the boundary condition

(34.9) $$P^*(I+\Omega(\sigma))z_\sigma^0 = Q^*(I+\Omega(\tau))z_\tau^0.$$

If $\mathcal{R}(Q-T(\tau,\sigma)P)$ is closed in C, this condition is both necessary and sufficient.

Proof: Proceeding as in the proof of Theorem 34.1, relation (34.7) is equivalent to

$$T(\tau,\sigma)\,p - q + K(\tau,\sigma)h \in \mathcal{R}(Q-T(\tau,\sigma)P).$$

Therefore, it is necessary, and under the closure hypothesis sufficient, that

$$T(\tau,\sigma)p - q + K(\tau,\sigma)h \in \{\mathfrak{N}(Q^*-P^*T^*(\sigma,\tau))\}^\perp ;$$

that is; for any $\psi \in B_0$ such that $Q^*\psi - P^*T^*(\sigma,\tau)\psi = 0$,

(34.10) $$0 = \langle \psi, T(\tau,\sigma)p - q + K(\tau,\sigma)h\rangle$$

$$= \langle T^*(\sigma,\tau)\psi,p\rangle - \langle\psi,q\rangle + \int_\sigma^\tau (K^*(\sigma,\tau)\psi)(s)h(s)ds$$

$$= \langle (I+\Omega(\sigma))\tilde{T}(\sigma,\tau)(I+\Omega(\tau))^{-1}\psi,p\rangle - \langle\psi,q\rangle - \int_\sigma^\tau (T^*(\xi,\tau)\psi)(0^-)h(\xi)d\xi .$$

The condition $Q^*\psi - P^*T^*(\sigma,\tau)\psi = 0$ is equivalent to

$$0 = Q^*\psi - P^*(I+\Omega(\sigma))\tilde{T}(\sigma,\tau)(I+\Omega(\tau))^{-1}\psi$$

If $z_\tau^0 = (I+\Omega(\tau))^{-1}\psi$, then the existence of a $\psi \in \mathfrak{N}(Q^*-P^*T^*(\sigma,\tau))$ is equivalent to the existence of a solution of the adjoint equation (34.4) on $[\sigma-r,\tau-r]$ satisfying (34.9). Using $z_\tau^0 = (I+\Omega(\tau))^{-1}\psi$, $z_\sigma^0 = \tilde{T}(\sigma,\tau)z_\tau^0$ and the same argument as in the last part of the proof of Theorem 34.1, the relation (34.8) is obtained from (34.10). This proves the theorem.

Corollary 34.1. Suppose $h(t)$, $L(t,\varphi)$ in (34.1) are periodic in t of period $\omega > 0$. The necessary and sufficient condition that there exist ω-periodic solutions of (34.1) is

$$\int_0^\omega z(\alpha)h(\alpha)d\alpha = 0 \tag{34.11}$$

for all ω-periodic solutions z of the adjoint equation (34.4).

Proof: Take $\sigma = 0$, $\tau = \omega$, $V = C$, $\gamma = 0$, $M = -N = I$ in Theorem 34.1. Some iterate of $T(\omega,0)$ is compact and, therefore, $I - T(\omega,0)$ has closed range. The corollary now follows from Theorem 34.1.

As another application, consider the differential difference equation

$$\dot{x}(t) = A(t)x(t) + B(t)x(t-r) + h(t), \qquad \sigma \le t \le \tau \tag{34.12}$$

where $\tau - \sigma \ge r$ and suppose M,N are $n \times n$ matrices. The problem is to find a $\varphi \in C$ and a solution of (34.12) such that

$$(34.13) \qquad x_\sigma = M\varphi, \quad x_\tau = N\varphi,$$

where $(M\varphi)(\theta) = M\varphi(\theta)$, $(N\varphi)(\theta) = N\varphi(\theta)$, $-r \leq \theta \leq 0$.

The function η in the definition of L is given by $\eta(t,\theta) = 0$ for $\theta \geq 0$, $= -A(t)$ for $-r < \theta < 0$, $= -A(t) - B(t)$ for $\theta \leq -r$. Therefore,

$$(34.14) \qquad (\Omega(t)\psi)(\theta) = -\int_\theta^0 \psi(\alpha)A(t+\alpha)d\alpha, \qquad -r \leq \theta \leq 0.$$

Also, $M^*\psi(\theta) = \psi(\theta)M$, $N^*\psi(\theta) = \psi(\theta)N$ for $\psi \in B_0$. According to Theorem 34.2 with $p = q = 0$, a necessary condition that (34.12), (34.13) have a solution is that

$$(34.15) \qquad \int_\sigma^\tau z(\xi)h(\xi)d\xi = 0$$

for all those solutions z of the adjoint equation

$$(34.16) \qquad \dot{z}(t) + z(t)A(t) + z(t+r)B(t+r) = 0$$

which satisfy the boundary conditions (34.9); that is,

$$(34.17) \qquad [z_\sigma(\theta) - \int_\theta^0 z_\sigma(\alpha)A(\sigma+\alpha)d\alpha]M = [z_\tau(\theta) - \int_\theta^0 z_\tau(\alpha)A(\tau+\alpha)d\alpha]N.$$

If $\mathscr{R}[N-T(\tau,\sigma)M]$ is closed, then condition (34.15) is also sufficient.

If N is nonsingular, then $\mathscr{R}[N-T(\tau,\sigma)M] = \mathscr{R}[I-T(\tau,\sigma)MN^{-1}]$ and $T(\tau,\sigma)$ compact implies this set is closed. On the other hand, if N is singular, then this range may not be closed.

Under additional hypotheses on M,N, the boundary condition (34.17) can be written in a simpler form. If M,N commute with $A(t)$ for each t and we make the further nonrestrictive condition that $A(t)$ is periodic of period $\tau - \sigma$, then (34.17) is equivalent to

$$u(\theta) - \int_{\theta}^{0} u(\alpha)A(\sigma+\alpha)d\alpha = 0$$

$$u(\theta) = z_{\sigma}(\theta)M - z_{\tau}(\theta)N, \qquad -r \leq \theta \leq 0.$$

But this Volterra equation has the unique solution $u(\theta) \equiv 0$; that is, the boundary condition is

$$z_{\sigma}(\theta)M = z_{\tau}(\theta)N, \qquad -r \leq \theta \leq 0. \tag{34.18}$$

It is instructive to analyze in more detail the meaning of the boundary conditions (34.5) for the adjoint equation (34.4). In particular, what is the meaning of $(I+\Omega(\sigma))^{-1}$? Suppose $z(s)$ is a solution of (34.4) for $s \leq \tau$, $z(s) = 0$ for $s \geq \tau + r$ and $z^{o}_{\tau+r} = \psi \in B_0$. Then

$$z(s) + \int_{s}^{\tau} z(\alpha)\eta(\alpha,s-\alpha)d\alpha = z(\tau) + \int_{\tau}^{\infty} z(\alpha)\eta(\alpha,\tau-\alpha)d\alpha - \int_{\tau}^{\infty} z(\alpha)\eta(\alpha,s-\alpha)d\alpha$$

and for $-r \leq \theta \leq 0$

$$z_{\tau}(\theta) + \int_{\theta}^{0} z_{\tau}(\alpha)\eta(\tau+\alpha,\theta-\alpha)d\alpha = \Gamma(\tau+r,z_{t+r})(\theta) \tag{34.19}$$

$$\Gamma(\tau,\psi)(\theta) = \psi(-r) + \int_{-r}^{0} \psi(\alpha)[\eta(\tau-\alpha,-r-\alpha) - \eta(\tau-\alpha,\theta-r-\alpha)]d\alpha, \quad -r \leq \theta \leq 0. \tag{34.20}$$

Consequently,

$$(I+\Omega(\tau))z^{o}_{t} = \Gamma^{o}(\tau+r,\psi) \tag{34.21}$$

and $(I+\Omega(\tau))^{-1}\Gamma^{o}(\tau+r,\psi)$ represents the solution of the adjoint equation on $[\tau-r,\tau]$ corresponding to the initial value ψ on $[\tau,\tau+r]$. The boundary conditions (34.5) therefore states that $\Gamma^{o}(\tau+r,\psi) = N^{*}\delta$ and $\Gamma^{o}(\tau+r,\psi)$ is specified through N^{*} and elements $\delta \in V^{*}$ rather than being specified through the function Γ and elements $\psi \in B_0$ corresponding to initial values of solutions of the adjoint on $[\tau,\tau+r]$.

It is a natural question to ask and certainly would be useful to be able to specify the boundary condition in terms of the initial value on $[\tau,\tau+r]$. This involves determining some appropriate pseudo inverse of the operator $\Gamma^0(\tau+r,\cdot)$. To illustrate the problems involved, let us consider the differential difference equation

$$\dot{x}(t) = Ax(t) + Bx(t-r) \tag{34.22}$$

where A,B are constant $n \times n$ matrices. For this case $\Omega(\sigma) = \Omega$ for every σ and, as in the discussion of (34.12),

$$(\Omega\psi)(\theta) = -\int_\theta^0 \psi(\alpha)d\alpha A = \int_0^\theta \psi(\alpha)d\alpha A \tag{34.23}$$

for $-r \leq \theta \leq 0$. Also, a few elementary calculations give

$$((I+\Omega)^{-1}\psi)(\theta) = \psi(\theta) - \int_0^\theta \psi(s)e^{-A(\theta-s)}Ads, \quad -r \leq \theta \leq 0. \tag{34.24}$$

Suppose we consider the following boundary value problem for (34.22). Let $V = W \times E^n$ where W is a Banach space and with elements γ of V given as $\gamma = (\delta_1,\delta_2)$, $\delta_1 \in W$, $\delta_2 \in E^n$. Let $N\colon C \to V$ be defined by $N\varphi = (0,\varphi(0))$ and $M\colon C \to V$ defined by $M\varphi = (M_1\varphi,0)$ where $M_1\colon C \to W$ is a linear operator with domain dense in C. For a fixed element $\gamma = (\varphi_0,b) \in W$ the boundary condition $Mx_0 + Nx_\tau = \gamma$ for (34.1) is equivalent to

$$M_1x_0 = \varphi_0, \quad x(\tau) = b. \tag{34.25}$$

The operator N^* is defined by

$$\langle N^*\delta,\varphi\rangle = \langle \delta, N\varphi\rangle_V = \delta_2\varphi(0) \quad \text{for every} \quad \delta = (\delta_1,\delta_2) \in V^*, \quad \varphi \in C.$$

Since $\langle N^*\delta,\varphi\rangle = \int_{-r}^{0} [d_\theta N^*\delta(\theta)]\varphi(\theta)$ this implies that

$$(34.26) \qquad N^*\delta(\theta) = \begin{cases} 0 & \theta = 0 \\ \delta_2 & -r \le \theta < 0 \end{cases}.$$

Furthermore, from (34.24),

$$(34.27) \qquad (I+\Omega)^{-1}N^*\delta(\theta) = \begin{cases} 0 & \theta = 0 \\ \delta_2 e^{-A\theta}, & -r \le \theta < 0 \end{cases}.$$

From (34.20), $\Gamma(\tau,\psi) = \Gamma(\psi)$ for all τ and

$$(34.28) \qquad \Gamma^0\psi(\theta) = \psi(-r) + \int_{\theta}^{0} \psi(\alpha)B\,d\alpha, \quad -r \le \theta \le 0.$$

The question of being able to specify the boundary condition on $[\tau,\tau+r]$ is equivalent to the following: For any $\delta = (\delta_1,\delta_2) \in V^*$, does there exist a $\psi = \psi(\delta)$ such that $\Gamma^0\psi = N^*\delta$, that is

$$(34.29) \qquad \psi(-r) + \int_{\theta}^{0} \psi(\alpha)B\,d\alpha = \delta_2, \quad -r \le \theta < 0.$$

This later equation implies (by differentiation with respect to θ) that $\psi(\theta)B = 0$ for all $\theta \in (-r,0)$. Therefore, ψ is uniquely determined if and only if B is nonsingular. In any case, the function

$$\psi(\theta) = \begin{cases} \delta_2 & \theta = -r \\ 0 & -r < \theta \leq 0 \end{cases}$$

is such that $\Gamma^{o}\psi = N^{*}\delta$.

A convenient choice for ψ for a more general boundary value problem remains unclear.

35. LINEAR PERIODIC SYSTEMS. GENERAL THEORY

Suppose $L: R \times C \to R^n$ is the function defined in (32.1), satisfies the conditions of Section 32 and, in addition, $L(t+\omega,\varphi) = L(t,\varphi)$ for all t,φ. In this section we consider the system

$$\dot{x}(t) = L(t,x_t) \tag{35.1}$$

and the extent to which a Floquet theory for (35.1) exists.

For any $s \in R$, $\varphi \in C$, there is a solution $x = x(s,\varphi)$ of (35.1) defined on $[s,\infty)$ and $x_t(s,\varphi)$ is continuous in t,s,φ. As usual, let $T(t,s)\varphi = x_t(s,\varphi)$ for all $t \geq s$, $\varphi \in C$. The operator $T(t,s)$ always satisfies $T(t,s)T(s,\tau) = T(t,\tau)$ for all $t \geq s \geq \tau$ and periodicity of (35.1) implies $T(t+\omega,s) = T(t,s)T(s+\omega,s)$ for all $t \geq s$. Let $U: C \to C$ be defined by

$$U\varphi = T(\omega,0)\varphi. \tag{35.2}$$

Let us suppose for the moment that $\omega \geq r$. This latter assumption implies that the image of any bounded set in C under U is relatively compact. Therefore, U is completely continuous and the spectrum $\sigma(U)$ of U is at most countable, is a compact subset of the complex plane with the only possible accumulation point being zero and $\mu \neq 0$ implies μ is in the point spectrum $P\sigma(U)$ of U, that is, there is a $\varphi \neq 0$ in C such that $U\varphi = \mu\varphi$. Any $\mu \neq 0$ in $P\sigma(U)$ is called a <u>characteristic multiplier</u> of (35.1) and any λ for which $\mu = e^{\lambda\omega}$ is called a <u>characteristic exponent</u> of (35.1).

<u>Lemma 35.1.</u> $\mu = e^{\lambda\omega}$ is a characteristic multiplier of (35.1) if and only if there is a $\varphi \neq 0$ in C such that $T(t+\omega,0)\varphi = \mu T(t,0)\varphi$ for all $t \geq 0$

<u>Proof:</u> If $\mu \in P\sigma(U)$, then there is a $\varphi \neq 0$ in C such that $U\varphi = \mu\varphi$. Periodicity of (35.1) implies $T(t+\omega,0)\varphi = \mu T(t,0)\varphi$ for all t. The converse is trivial.

Since U is completely continuous, for any characteristic multiplier μ

of (35.1) there are two closed subspaces E_μ, K_μ of C such that the following properties hold:

(i) E_μ is finite dimensional;

(ii) $E_\mu \oplus K_\mu = C$;

(iii) $UE_\mu \subset E_\mu$, $UK_\mu \subset K_\mu$;

(iv) $\sigma(U|E_\mu) = \{\mu\}$, $\sigma(U|K_\mu) = \sigma(U) - \{\mu\}$.

Let $\varphi_1,\ldots,\varphi_{d_\mu}$ be a basis for E_μ, $\Phi = (\varphi_1,\ldots,\varphi_{d_\mu})$. Since $UE_\mu \subset E_\mu$, there is a $d_\mu \times d_\mu$ matrix $M = M_\mu$ such that $U\Phi = \Phi M$ and property (iv) implies the only eigenvalue of M is $\mu \neq 0$. Therefore, there is a $d_\mu \times d_\mu$ matrix $B = B_\mu$ such that $B = \omega^{-1} \ln M$. Define the vector $P(t)$ with elements in C by $P(t) = T(t,0)\Phi e^{-Bt}$. Then, for $t \geq 0$,

$$\begin{aligned} P(t+\omega) &= T(t+\omega,0)\Phi e^{-B(t+\omega)} = T(t,0)T(\omega,0)\Phi e^{-B\omega}e^{-Bt} \\ &= T(t,0)U\Phi e^{-B\omega}e^{-Bt} = T(t,0)\Phi M e^{-B\omega}e^{-Bt} = P(t); \end{aligned}$$

that is, $P(t)$ is periodic of period ω. Thus, $T(t,0)\Phi = P(t)e^{Bt}$, $t \geq 0$. Extend the definition $P(t)$ for t in $(-\infty,\infty)$ in the following way. If $t < 0$, there is an integer k such that $t + k\omega \geq 0$ and we let $P(t) = P(t+k\omega)$. The function $x_t(0,\varphi) = T(t,0)\Phi = P(t)e^{Bt}$ is then well defined for $-\infty < t < \infty$ and it is easily seen that each column of this matrix is a solution of (35.1) on $(-\infty,\infty)$. We, therefore, have

<u>Lemma 35.2</u>. If μ is a characteristic multiplier of (32.1) and Φ is a basis for E_μ of dimension d_μ, then there are $d_\mu \times d_\mu$ matrices B, $\sigma(e^{B\omega}) = \{\mu\}$, and $P(t+\omega) = P(t)$, $t \in (-\infty,\infty)$ if $\varphi = \Phi b$, then $x_t(\varphi)$ is defined for $t \in (-\infty,\infty)$ and

$$x_t(0,\varphi) = P(t)e^{Bt}b, \quad t \in (-\infty,\infty).$$

Therefore, in particular, $\mu = e^{\lambda\omega}$ is a characteristic multiplier of (35.1) if and only if there is a nonzero solution of (35.1) of the form

$$x(t) = P(t)e^{\lambda t}$$

where $P(t+\omega) = P(t)$.

Since $x_t(0,\varphi)(\theta) = x(0,\varphi)(t+\theta) = x_{t+\theta}(0,\varphi)(0)$, $-r \leq \theta \leq 0$, and $\varphi \in E_\mu$ it follows that $P(t)(\theta) = P(t+\theta)(0)e^{B\theta}$, $-r \leq \theta \leq 0$. Therefore, if we let $\tilde{P}(t+\theta) = P(t+\theta)(0)$, then $\Phi(\theta) = \tilde{P}(\theta)e^{B\theta}$ and

$$x(0,\varphi)(t) = \tilde{P}(t)e^{Bt}b, \qquad t \in (-\infty,\infty),\ \varphi = \Phi b.$$

Therefore, the solutions of (35.1) with initial value in E_μ are of the Floquet-type; namely, if $\mu = e^{\lambda\omega}$ the solutions are of the form $e^{\lambda t}$ times a polynomial in t with coefficients periodic in t of period ω.

We also need the following remark: If $T(t,0)\Phi b = 0$ for any t, then $b = 0$. In fact, if there is a t such that $T(t,0)\Phi b = 0$,

$$0 = T(m\omega,0)\Phi b = U^m\Phi b = M^m\Phi b.$$

Since the eigenvalue of M^m is $\mu^m \neq 0$, the result follows immediately.

We have defined the characteristic multipliers of (35.1) in terms of the period map U of (35.1) starting with the initial time 0. To justify the terminology it is necessary to show that the multipliers do not depend upon the starting time. We prove much more than this.

For any s in R, let $U(s) = T(s+\omega,s)$. As before, for any $\mu \neq 0$, $\mu \in \sigma(U(s))$, there exist two closed subspaces $E_\mu(s)$, $K_\mu(s)$ of C such that properties (i) - (iv) hold with the appropriate change in notation. Let $\Phi(s)$ be a basis for $E_\mu(s)$, $U(s)\Phi(s) = \Phi(s)M(s)$, $\sigma(M(s)) = \{\mu\}$. As for the case $s = 0$, one can define $T(t,s)\Phi(s)$ for all $t \in (-\infty,\infty)$. For any real number τ

$$\begin{aligned} U(\tau)T(\tau,s)\Phi(s) &= T(\tau+\omega,\tau)T(\tau,s)\Phi(s) \\ &= T(\tau+\omega,s)\Phi(s) \\ &= T(\tau,s)T(s+\omega,s)\Phi(s) \\ &= T(\tau,s)\Phi(s)M(s). \end{aligned}$$

If $M = \mu I - Q$, then Q is nilpotent and

$$[\mu I-U(\tau)]T(\tau,s)\Phi(s) = T(\tau,s)\Phi(s)M.$$

Since $T(t,s)\Phi(s)b = 0$ implies $b = 0$, it follows that μ is in $\sigma(U(\tau))$ and the dimension of $E_\mu(\tau)$ is at least as large as the dimension of $E_\mu(s)$. Since one can reverse the role of s and τ, we obtain the following

Lemma 35.3. The characteristic multipliers of (35.1) are independent of the starting time and if $\Phi(s)$ is a basis for $E_\mu(s)$, then $T(t,s)\Phi(s)$ is a basis for $E_\mu(t)$ for any t in R.

If $0 < \omega < r$ and U is defined in the same way as $U = T(\omega,0)$, then there is an integer $m > 0$ such that $m\omega \geq r$. Since $U^m = T(m\omega,0)$, the operator U^m is compact. It then follows from Reisz-Nagy (p. 424) that the nonzero spectrum of U is only point spectrum, is a compact set in the complex plane with the only possible accumulation point being zero and for any $\mu \neq 0$ in $\sigma(U)$, the space C can be decomposed as above.

The next result relates the Floquet representation in each eigenspace to the solution $x_t(\varphi,s)$ for an arbitrary $\varphi \in C$. Suppose $\sigma(U) = \{0\} \cup \{\mu_m\}$ where $\{\mu_m\}$ is either finite or countable and each $\mu_m \neq 0$. Let $P_m(s)\colon C \to E_{\mu_m}(s)$, $I - P_m(s)\colon C \to K_{\mu_m}(s)$, be the projection induced by $E_{\mu_m}(s)$, $K_{\mu_m}(s)$ defined above and satisfying (i) - (iv).

Theorem 35.1. Suppose φ is a given element of C and the notation is as described above. If α is an arbitrary real number, then there are constants $\beta = \beta(\alpha) > 0$, $M = M(\alpha) > 0$ such that

$$|x_t(x,\varphi) - \sum_{|\mu_n| \geq \exp \alpha\omega} x_t(s,P_n(s)\varphi)|$$

$$\leq Me^{(\alpha-\beta)(t-s)}|\varphi|, \quad t \geq s.$$

Proof: For any integer k, one has

$$C = E_{\mu_1}(s) \oplus \cdots \oplus E_{\mu_k}(s) \oplus F_k(s)$$

for any set $\mu_1,\ldots,\mu_k$ of nonzero elements of $\sigma(U(s))$, where each $E_{\mu_j}(s)$ as well as $F_k(s)$ are invariant under $U(s)$ and

$$\sigma(U(s)|F_k(s)) = \sigma(U(s))\backslash\{\mu_1,\ldots,\mu_k\}.$$

Order the μ_n so that $|\mu_1| \geq |\mu_2| \geq \cdots \geq 0$ and, for any α, let $k = k(\alpha)$ be the integer satisfying $|\mu_k| \geq \exp \alpha\omega$ and $|\mu_{k+1}| < \exp \alpha\omega$. If

$$R_k(\varphi) = \varphi - \sum_{j=1}^{k} P_j(\varphi)$$

then $R_k(\varphi) \in F_k(s)$, $x_{s+m\omega}(s,R_k\varphi) \in F_k(s)$ for all $m = 0,1,2,\ldots,\varphi \in C$. If $e^{\gamma\omega} = |\mu_{k+1}|$, then the spectral radius of $U_1 \overset{\text{def}}{=} U(s)|F_k(s)$ is $e^{\gamma\omega}$. Therefore, (Reisz-Nagy, p. 425), $\lim_{n\to\infty} |U_1^n|^{1/n} = e^{\gamma\omega}$. The proof is completed exactly as we did for the case when (35.1) was independent of t (see Section 22).

Corollary 35.1. The solution $x = 0$ of (35.1) is uniformly asymptotically stable if and only if all characteristic multipliers of (35.1) have modulii less than 1.

Proof: The if part follows from Theorem 35.1 with $\alpha = 0$. The only if part is a consequence of the Floquet representation associated with any characteristic multiplier.

In a similar manner, one obtains

<u>Corollary 35.2</u>. The solution $x = 0$ of (35.1) is uniformly stable if and only if all characteristic multipliers of (35.1) have modulii ≤ 1 and if μ is a multiplier with $|\mu| = 1$, then all solutions of (35.1) with initial value in E_μ are bounded.

<u>Corollary 35.3</u>. If $\Sigma_m P_m(s)\varphi$ converges and $R\varphi = \varphi - \Sigma_m P_m(s)\varphi$, then, for any real number α,

$$e^{\alpha t}|x_t(s,R\varphi)| \to 0 \quad \text{as} \quad t \to \infty.$$

Our next objective is to show how, in particular cases, one can determine the characteristic multipliers of (35.1). Consider the system

$$\dot{x}(t) + \sum_{j=0}^{m} B_j(t)x(t-k_j\omega) = 0 \tag{35.2}$$

where $B_j(t+\omega) = B_j(t)$ and each $k_j \geq 0$ is an integer. From Lemma (35.2), $\mu = e^{\lambda\omega}$ is a characteristic multiplier of (35.2) if and only if there is a nonzero n-vector $v(t) = v(t+\omega)$ such that $x(t) = v(t)e^{\lambda t}$ satisfies (35.2). Therefore, since the k_j are integers, this implies that

(35.3)

$$\text{(a)} \qquad \dot{v}(t) + \left[\lambda I + \sum_{j=0}^{m} B_j(t)e^{-k_j\omega\lambda}\right]v(t) = 0,$$

$$\text{(b)} \qquad v(t+\omega) = v(t).$$

If $V(t,\lambda)$, $V(0,\lambda) = I$, is the principal matrix solution of (35.3a) then $v(t) = V(t,\lambda)v(0)$ and the initial value $v(0) \neq 0$ must be chosen in such a way that (35.3b) is satisfied. Such a $v(0) \neq 0$ exists if and only if λ satisfies the "characteristic equation"

(35.4) $$\det [V(\omega,\lambda) - I] = 0.$$

In theory, at least, Equation (35.4) can be solved and one obtains, in particular, from Corollary 35.1 that all roots λ of (35.4) having negative real parts implies that the solution $x = 0$ of (35.2) is uniformly asymptotically stable.

For the case in which x in (35.2) is a scalar, one can obtain (35.4) explicitly in terms of the coefficients $B_j(t)$ in (35.2). In fact, for scalar equations, if we let $B_{j0} = \omega^{-1} \int_0^\omega B_j(t)dt$, then, since the λ are determined only up to a multiple of $2\pi/\omega$, the condition $V(\omega,\lambda) - I = 0$ will be satisfied if and only if

(35.5) $$\lambda + \sum_{j=0}^{m} B_{j0} \exp (-k_j\omega\lambda) = 0.$$

As an example, consider the scalar equation

(35.6) $$\dot{y}(t) = (\sin t)y(t-2\pi).$$

From (35.5) the number $\mu = e^{2\pi\lambda}$ is a characteristic multiplier of (35.6) if and only if $\lambda = 0$; that is, there is only one characteristic multiplier $\mu = 1$ of (35.6). Furthermore, one easily shows that a basis for E_1 is $p(t) = e^{-\cos t}$, $-2\pi \leq t \leq 0$. Finally, if P is the projection defined by the above decomposition of C induced by the multiplier $\mu = 1$, then $x_t(0,\varphi-P\varphi) \to 0$ as $t \to \infty$ faster than any exponential.

36. DECOMPOSITION OF LINEAR PERIODIC SYSTEMS

In this section, we consider the linear periodic system (35.1) and the problem of the decomposition of the space C using the generalized eigenspaces of the characteristic multipliers of (35.1). The decomposition is given explicitly through a bilinear form similar to the one for the autonomous equation in Section 21. The results are easy consequences of the adjoint theory of Section 33.

For a given σ, if $U(\sigma) = T(\sigma+\omega,\sigma)$, $\tilde{U}(\sigma) = \tilde{T}(\sigma,\sigma+\omega)$, then $U^*(\sigma) = (I+\Omega(\sigma))\tilde{U}(\sigma)(I+\Omega(\sigma))^{-1}$ from Theorem 33.1 and the fact that $\Omega(\sigma+\omega) = \Omega(\sigma)$ in (33.6). We have already seen in Section 35 that, for some integer m, $E_\mu(\sigma) = \mathfrak{N}(\mu I-U(\sigma))^m$, $K_\mu(\sigma) = \mathscr{R}(\mu I-U(\sigma))^m$ and

$$C = \mathfrak{N}(\mu I-U(\sigma))^m \oplus \mathscr{R}(\mu I-U(\sigma))^m. \tag{36.1}$$

Since $\tilde{U}(\sigma)$ is similar to $U^*(\sigma)$, the spectrum of $\tilde{U}(s)$ and $U(s)$ are the same. Furthermore,

$$B_0 = \mathfrak{N}(\mu I-\tilde{U}(\sigma))^m \oplus \mathscr{R}(\mu I-\tilde{U}(\sigma))^m, \tag{36.2}$$

with $\dim \mathfrak{N}(\mu I-U(\sigma))^m = \dim \mathfrak{N}(\mu I-\tilde{U}(\sigma))^m < \infty$.

To give an explicit characterization of the decompositions (36.1), (36.2), introduce the bilinear form

$$[\psi|\varphi]_t = \langle (I+\Omega(t))\psi,\varphi\rangle, \qquad \psi \in B_0,\ \varphi \in C. \tag{36.3}$$

Using Theorem 33.1, we have

$$[\psi|T(t,\sigma)\varphi]_t = [\tilde{T}(\sigma,t)\psi|\varphi]_\sigma, \quad \psi \in B_0,\ \varphi \in C,\ t \geq \sigma, \tag{36.4}$$

for the general linear system. For the periodic system (35.1) $\Omega(t+\omega) = \Omega(t)$ for all t and therefore

$$[\psi|U(\sigma)\varphi]_\sigma = [\tilde{U}(\sigma)\psi|\varphi]_\sigma, \quad \psi \in B_0, \ \varphi \in C; \tag{36.5}$$

that is, $U(\sigma), \tilde{U}(\sigma)$ are adjoints relative to the nonsingular bilinear form $[\psi|\varphi]_\sigma$.

Now let $\Psi^\mu = \mathrm{col}(\psi_1^\mu, \ldots, \psi_d^\mu)$, $\Phi^\mu = (\varphi_1^\mu, \ldots, \varphi_d^\mu)$ be bases for $\mathfrak{N}(\mu I - \tilde{U}(\sigma))^m$, $\mathfrak{N}(\mu I - U(\sigma))^m$, respectively. Then

$$\mathscr{R}(\mu I - U(\sigma))^m = \{\varphi \in C: [\Psi^\mu|\varphi]_\sigma = 0\} \tag{36.6}$$

$$\mathscr{R}(\mu I - \tilde{U}(\sigma))^m = \{\psi \in B_0: [\psi|\Phi^\mu]_\sigma = 0\}. \tag{36.7}$$

In fact, if $\varphi = (\mu I - U(\sigma))^m \nu$ for some $\nu \in C$, then $[\Psi^\mu|\varphi]_\sigma = [\Psi^\mu|(\mu I - U(\sigma))^m \nu]_\sigma = [(\mu I - \tilde{U}(\sigma))^m \Psi^\mu|\nu]_\sigma = 0$. Conversely, if $[\Psi^\mu|\varphi]_\sigma = 0$ and $\varphi = \varphi_1 + \varphi_2$, $\varphi_1 \in \mathfrak{N}(\mu I - U(\sigma))^m$, $\varphi_2 \in \mathscr{R}(\mu I - U(\sigma))^m$, then $[\Psi^\mu|\varphi_2]_\sigma = 0$ and, thus, $[\Psi^\mu|\varphi_1]_\sigma = 0$. But, if $\psi \in \mathscr{R}(\mu I - \tilde{U}(\sigma))^m$; i.e., $\psi = (\mu I - \tilde{U}(\sigma))^m \mu$ for some μ in B_0, then $[\psi|\varphi_1]_\sigma = 0$. Therefore, relation (36.2) implies that $[\psi|\varphi_1]_\sigma = 0$ for all $\psi \in B_0$. The non-singularity of the bilinear form implies that $\varphi_1 = 0$. Therefore, $\varphi = \varphi_2$ and this proves (36.6). Relation (36.7) is proved in an analogous manner.

Relation (36.6) implies that the $d \times d$ matrix $[\Psi^\mu|\Phi^\mu]_\sigma$ is nonsingular and without any loss in generality can be chosen as the identity. Therefore, the decomposition (36.1), (36.2) can be written as

$$C = \{\varphi: \varphi = \Phi^\mu[\Psi^\mu|\varphi]_\sigma\} \oplus \{\varphi: [\Psi^\mu|\varphi]_\sigma = 0\} \tag{36.8}$$

$$B_0 = \{\psi: \psi = [\psi|\Phi^\mu]_\sigma \Psi^\mu\} \oplus \{\psi: [\psi|\Phi^\mu]_\sigma = 0\}. \tag{36.9}$$

Relations (36.8), (36.9) are sufficient for the applications, but some remakrs are in order to clarify the relationship between this decomposition and the one given in Section 21 for autonomous equations. The bilinear form in Section 21 is not the same one as used above and, in fact, could be singular. On the other hand, we proved in Section 21 that the corresponding matrix (Ψ^{μ},Φ^{μ}) for that bilinear form was nonsingular and thus could be used to decompose C. We now prove that the results in Section 21 are implied by the more general discussion above.

Let us introduce the bilinear form

$$(\psi,\varphi)_t = \psi(0)\varphi(0) + \int_{-r}^{0} d_\beta\{\int_0^r \psi(\xi)\eta(t+\xi,\beta-\xi)d\xi\}\varphi(\beta) \tag{36.10}$$

for $\varphi \in C$, ψ: $[0,r] \to E^{n*}$ of bounded variation. This reduces to the bilinear form discussed in Section 21 for the autonomous case. Let $S\psi$ be defined by $(S\psi)(\theta) = \psi(r+\theta)$, $-r \leq \theta \leq 0$. We assert that

$$\begin{aligned}(\psi,\varphi)_t &= -\langle (I+\Omega(t+r))S\psi, T(t+r,t)\varphi\rangle \\ &= -[S\psi|T(t+r,t)\varphi]_{t+r} = -[\tilde{T}(t,t+r)S\psi|\varphi]_t,\end{aligned} \tag{36.11}$$

for all $\varphi \in C$, ψ: $[0,r] \to E^{n*}$ such that $S\psi \in B_0$.

To prove (36.11), let $x_t = \varphi$, $x_{t+r} = T(t+r,t)\varphi$. Then

$$\begin{aligned}-\langle S\psi, x_{t+r}\rangle &= -\int_{-r}^{0} [d\psi(r+\theta)]x(t+r+\theta) \\ &= \psi(0)x(t) + \int_0^r \psi(\beta)\dot{x}(t+\beta)d\beta\end{aligned}$$

and

$$\begin{aligned}\langle -\Omega(t+r)S\psi, x_{t+r}\rangle &= -\int_0^r d_\xi[\int_0^r \psi(\beta)\eta(t+\beta,\xi-\beta)d\beta]x(t+\xi) \\ &= -\int_0^r \psi(\beta)[\int_0^r (d_\xi\eta(t+\beta,\xi-\beta))x(t+\xi)]d\beta \\ &= -\int_0^r \psi(\beta)[\dot{x}(t+\beta) - \int_{-r}^{0} d_\xi\eta(t+\beta,\xi-\beta)x(t+\xi)]d\beta\end{aligned}$$

$$= -\int_0^r \psi(\beta)\dot{x}(t+\beta)d\beta + \int_{-r}^0 d_\xi\{\int_0^r \psi(\beta)\eta(t+\beta,\xi-\beta)d\beta\}x(t+\xi).$$

The sum of these two expressions gives (36.11).

Relation (36.11) and Theorem 33.1 imply that

$$(36.12) \qquad (\psi,T(t,s)\varphi)_t = (S^{-1}\tilde{T}(s+r,t+r)S\psi,\varphi)_s.$$

For the autonomous case, relation (36.12) corresponds to formula (21.5).

Suppose μ is a characteristic multiplier of system (35.1), Φ^μ is a basis of the generalized eigenspace of $\mu I - U(\sigma)$ and $\bar{\Psi}^\mu$ is a basis for the generalized eigenspace of $\mu I - \tilde{U}(\sigma+r)$. The analogue of Lemma 35.3 for the "adjoint" equation implies that $\Psi^\mu \overset{\text{def}}{=} \tilde{T}(\sigma,\sigma+r)\bar{\Psi}^\mu$ is a basis for the generalized eigenspace of $\mu I - \tilde{U}(\sigma)$. Relation (36.11) gives

$$(S^{-1}\bar{\Psi}^\mu,\Phi^\mu)_\sigma = -[\Psi^\mu|\Phi^\mu]_\sigma$$

and, therefore, $(S^{-1}\bar{\Psi}^\mu,\Phi^\mu)_\sigma$ is nonsingular. It is also not difficult to show that $\Psi^\mu(\theta)$ and, therefore, $\bar{\Psi}^\mu(\theta)$ are absolutely continuous functions in θ except at $\theta = 0$. From the definition of $(\psi,\varphi)_\sigma$ in (36.10) and the properties of η, the value of $(\psi,\varphi)_\sigma$ does not depend on the value of ψ at $-r$. Therefore, $(S^{-1}\bar{\Psi}^\mu,\Phi^\mu)_\sigma$ does not depend upon the value of $\bar{\Psi}^\mu(\theta)$ at $\theta = 0$. Consequently, $\bar{\Psi}^\mu$ can be chosen to be a continuous function on $[-r,0]$.

In summary, the decomposition of C by a characteristic multiplier of (35.1) can be made using continuous solutions of the adjoint equation and the bilinear form $(\ ,\ )_t$ as well as by the bilinear form $[\ |\ ]_t$. To make the notation conform a little more with that in Section 21, let $C^* = \{\tilde{\psi}: [0,r] \to E^{n*}$ which are continuous$\}$. The above remarks imply that if $\tilde{\Psi}^\mu(\sigma) = (\tilde{\psi}_1^\mu,\ldots,\tilde{\psi}_d^\mu)$, $\psi_j^\mu \in C^*$, is a basis for the solutions of the adjoint equation on $[\sigma,\sigma+r]$ corresponding to the multiplier μ and $\Phi^\mu(\sigma)$ is a basis for the solutions of (35.1) on $[\sigma-r,\sigma]$ corresponding to the multiplier μ, then $(\tilde{\Psi}^\mu(\sigma),\Phi^\mu(\sigma))_\sigma$ is nonsingular. Without loss of generality, this matrix can be chosen to be the identity and we can write

$$C = \{\varphi\colon \varphi = \Phi^{\mu}(\sigma)(\tilde{\Psi}^{\mu}(\sigma),\varphi)_{\sigma}\} \oplus \{\varphi\colon (\tilde{\Psi}^{\mu}(\sigma),\varphi)_{\sigma} = 0\}. \tag{36.13}$$

In the remainder of this section $\tilde{\Psi}^{\mu}(\sigma)$, $\Phi^{\mu}(\sigma)$ will designate matrices chosen as above.

These remarks and relation (36.13) show in particular the validity of the decomposition used in Section 21 is a consequence of the general adjoint theory.

We now wish to obtain the same type of decomposition as above in the variation of constants formula for

$$\dot{x}(t) = L(t,x_t) + f(t), \tag{36.14}$$

where $L(t,\varphi)$ is the same function as in Section 32 and f is locally integrable on $(-\infty,\infty)$. The variation of constants formula for (36.14) is

$$x_t = T(t,t_0)x_{t_0} + \int_{t_0}^{t} T(t,s)X_0 f(s)ds, \qquad t \geq t_0, \tag{36.15}$$

where X_0 is the special $n \times n$ matrix function given by $X_0(\theta) = 0$, $-r \leq \theta < 0$, $X_0(0) = I$. the identity.

For any characteristic multiplier $\mu \neq 0$ of (35.1), let

$$x_t = x_t^{E_\mu(t)} + x_t^{K_\mu(t)} \quad \text{where} \quad x_t^{E_\mu(t)} \in E_\mu(t),\ x_t^{K_\mu(t)} \in K_\mu(t)$$

for any $t \geq t_0$. To find the integral equation for the components of x_t, let

$$T(t,s)X_0 = T(t,s)X_0^{E_\mu(s)} + T(t,s)X_0^{K_\mu(s)}. \tag{36.16}$$

To show that the latter object has meaning, observe that each column of $T(t,s)X_0$ belongs to C for $t \geq s + r$ and also

$$T(t,s) = T(t,s+r)T(s+r,s)X_0, \qquad t \geq s + r.$$

Therefore, each column of $T(s+r,s)X_0$ can be decomposed into its components according to the decomposition $E_\mu(s+r) \oplus K_\mu(s+r)$. Since $T(s+r,s)$ is a homeomorphism on $E_\mu(s+r)$, this allows one to define in a unique manner an $n \times n$ matrix $X_0^{E_\mu(0)}$ whose columns are in $E_\mu(s)$ so that

$$[T(s+r,s)X_0]^{E_\mu(s+r)} = T(s+r,s)X_0^{E_\mu(s)}.$$

If one defines

$$X_0^{K_\mu(s)} = X_0 - X_0^{E_\mu(s)},$$

then

$$[T(s+r,r)X_0]^{K_\mu(s+r)} = T(s+r,s)X_0^{K_\mu(s)}.$$

This justifies (36.16).

We have seen above that the decomposition of C by characteristic multipliers could be explicitly performed either with the bilinear form $[\ |\ ]_t$ in (36.3) or the bilinear form $(\ ,\ )_t$ in (36.10). In keeping with Section 21, we prefer to use $(\ ,\ )_t$.

If y is a solution of the "adjoint" equation (32.4), we let $y^t = S^{-1}y_{t+r}$. The following lemma is needed and a proof is supplied by a direct computation in (36.13) and (32.4).

LEMMA 36.1. If y is a solution of (32.4) on $(-\infty,\infty)$ and x is a solution of

(36.13) for $t \geq t_0$, then

$$\frac{d}{dt}(y^t, x_t)_t = y(t)f(t), \qquad t \geq \sigma,$$

and

$$(y^t, x_t)_t = (y^{t_0}, x_{t_0})_{t_0} + \int_{t_0}^{t} y(s)f(s)ds.$$

<u>Theorem 36.1</u>. If x is a solution of (36.13) for $t \geq t_0$ and $\mu \neq 0$, is a characteristic multiplier of (35.1) which decomposes C for any $s \in (-\infty, \infty)$ as $C = E_\mu(s) \oplus K_\mu(s)$ with $E_\mu(s), K_\mu(s)$ as in Section 32, then x_t satisfies the integral equations

$$(36.17) \qquad \begin{aligned} x_t^{E_\mu(t)} &= T(t,t_0)x_{t_0}^{E_\mu(t_0)} + \int_{t_0}^{t} T(t,s)X_0^{E_\mu(s)} f(s)ds, \\ x_t^{K_\mu(t)} &= T(t,t_0)x_{t_0}^{K_\mu(t_0)} + \int_{t_0}^{t} T(t,s)X_0^{K_\mu(s)} f(s)ds, \quad t \geq t_0. \end{aligned}$$

<u>Proof:</u> Suppose $\Phi(0) = \Phi^\mu(0)$, $\tilde{\Psi}(0) = \tilde{\Psi}^\mu(0)$ are bases chosen as stated above before (36.13) with $(\tilde{\Psi}(0), \Phi(0))_0 = I$, the identity. Let $\Phi(t) = T(t,0)\Phi(0)$ and let $\tilde{\Psi}(t)$ be the matrix solution of the adjoint equation on $[t, t+r]$ with initial value $\tilde{\Psi}(0)$ on $[0,r]$. Lemma 35.3 and the corresponding generalization for the adjoint equation imply that $\Phi(t)$ is a basis for the solutions of (35.1) on $[t-r, t]$ corresponding to the multiplier μ and $\tilde{\Psi}(t)$ is a basis for the solutions of the adjoint equation on $[t, t+r]$ corresponding to the multiplier μ. From Lemma 36.1, $(y^t, x_t)_t$ = constant and thus $(\tilde{\Psi}(t), \Phi(t))_t = I$ for all $t \in (-\infty, \infty)$. Furthermore, $x_t^{E_\mu(t)} = \Phi(t)(\tilde{\Psi}(t), x_t)_t$ for $t \geq t_0$ and Lemma 36.1 yield

$$
\begin{aligned}
x_t^{E_\mu(t)} &= \Phi(t)(\tilde{\Psi}(t),x_t)_t \\
&= \Phi(t)(\tilde{\Psi}(t_0),x_{t_0})_{t_0} + \Phi(t)\int_{t_0}^{t} \tilde{\Psi}(s)(0)f(s)ds \\
&= T(t,t_0)\Phi(t_0)(\tilde{\Psi}(t_0),x_{t_0})_{t_0} + \int_{t_0}^{t} T(t,s)\Phi(s)\tilde{\Psi}(s)(0)f(s)ds \\
&= T(t,t_0)x_{t_0}^{E_\mu(t_0)} + \int_{t_0}^{t} T(t,s)X_0^{E_\mu(s)} f(s)ds
\end{aligned}
$$

since $X_0^{E_\mu(s)} = \Phi(s)(\tilde{\Psi}(s),X_0)_s = \Phi(s)\tilde{\Psi}(s)(0)$. Using the fact that $x_t^{K_\mu(t)} = x_t - x_t^{E_\mu(t)}$, one completes the proof of the theorem.

The first equation in (36.17) is equivalent to an ordinary differential equation. In fact, if $\Phi(t) = T(t)\Phi(0) = P(t)e^{Bt}$ and $\tilde{\Psi}(t) = e^{-Bt}P^*(t)$, $U\Phi = \Phi e^{B\omega}$, and

$$
(36.18) \qquad x_t = x_t^{E_\mu(t)} + x_t^{K_\mu(t)} = P(t)y(t) + x_t^{K_\mu(t)}
$$

then (36.17) is equivalent to the system

$$
(36.19) \qquad
\begin{aligned}
\dot{y}(t) &= By + P^*(t)(0)f(t) \\
x_t^{K_\mu(t)} &= T(t,t_0)x_{t_0}^{K_\mu(t_0)} + \int_{t_0}^{t} T(t,s)X_0^{K_\mu(s)} f(s)ds, \quad t \geq t_0.
\end{aligned}
$$

System (36.19) is now in a form to permit the discussion of problems concerning the perturbation of Equation (35.1) in a manner very similar to that when (35.1) was autonomous. We do not devote any time to a detailed discussion of these questions since they proceed in a manner which is very analogous to ordinary differential equations. On the other hand, since we will discuss in some detail the neighborhood of a periodic orbit of an autonomous equation, it is necessary to have the following result for the case in which f in (36.13) is ω-periodic.

For the next lemma, we use the representation

$$L(t,\varphi) = \int_{-r}^{0} [d_\theta \eta(t,\theta)]\varphi(\theta)$$

for the function $L(t,\varphi)$.

Lemma 36.2. If $\mu = 1$ is a simple characteristic multiplier of system (35.1) [that is, $\dim \mathfrak{M}_{\mu=1}(U) = 1$], $p(t)$ is a nontrivial ω-periodic solution of (35.1) and $q(t)$ is a nontrivial ω-periodic solution of the adjoint equation and

$$J(t,p) \overset{\text{def}}{=} p(t) - \int_{0}^{\omega} [d_\theta \eta(t,\theta)]\theta p(t+\theta) \tag{36.20}$$

then

$$\int_{0}^{\omega} q(t)J(t)dt \neq 0. \tag{36.21}$$

Proof: Suppose $x(t)$ is a solution of the Equation (35.1). If $z(t) = x(t) + tp(t)$, then

$$\begin{aligned} \dot{z}(t) &= \dot{x}(t) + t\dot{p}(t) + p(t) \\ &= L(t,x_t) + tL(t,p_t) + p(t) \\ &= L(t,x_t + (t+\cdot)p_t) + p(t) - L(t,(\cdot)p_t) \end{aligned}$$

or

$$\dot{z}(t) = L(t,z_t) + J(t,p). \tag{36.22}$$

If (36.21) is not satisfied, then Corollary 34.1 implies there is a nontrivial periodic function $z(t)$ satisfying (36.22). Retracing the above steps, this implies there is a solution

$$x(t) = z(t) - tp(t)$$

of (35.1) where p, z are ω-periodic. This clearly implies $\dim \mathfrak{M}_{\mu=1}(U) \geq 2$ and contradicts the hypothesis that μ is simple.

Lemma 36.3. The equation (36.14) has an ω-periodic solution for an ω-periodic f if and only if

$$\int_0^\omega q(t)f(t)dt = 0$$

for all ω-periodic solutions q of the adjoint equation.

The proof of this lemma uses (36.19) and arguments similar to the ones used for the corresponding result for forced autonomous systems.

37. NONDEGENERATE PERIODIC ORBITS

Suppose $f: C \to R^n$ is continuous together with its Frechet derivative and consider the autonomous equation

$$\dot{x}(t) = f(x_t). \tag{37.1}$$

A periodic solution of (37.1) is a function $p(t) = p(t+\omega)$, $-\infty < t < \infty$, which satisfies (37.1) for $-\infty < t < \infty$. If p is a nonconstant periodic solution of (37.1), the orbit $\Gamma = \cup_t p_t$ of p is a closed curve and, conversely, any such orbit must correspond to a nonconstant periodic solution of (37.1).

If $p(t)$ is a nonconstant ω-periodic solution of (37.1), the linear variational equation relative to p is defined as

$$\dot{y}(t) = L(t, y_t) \tag{37.2}$$

where $L(t,\varphi) = f'(p_t)\varphi$ where $f'(\psi)$ is the Frechet derivative of f at ψ. The function $L(t,\varphi)$ has a representation

$$L(t,\varphi) = \int_{-r}^{0} [d_\theta \eta(t,\theta)]\varphi(\theta) \tag{37.3}$$

and $L(t+\omega,\varphi) = L(t,\varphi)$. Therefore, the linear variational equation is a linear periodic system. Furthermore, since $\dot{p}(t) = f(p_t)$ for $-\infty < t < \infty$, it follows that $\ddot{p}$ exists and $\ddot{p}(t) = f'(p_t)\dot{p}_t$. Therefore, $\dot{p}(t)$ is an ω-periodic solution of (37.2). Since we have assumed that $\dot{p}(t) \neq 0$, it follows that $\mu = 1$ is a characteristic multiplier of (37.2). We say the <u>periodic orbit</u> Γ <u>is nondegenerate</u> if the characteristic multiplier $\mu = 1$ is simple; that is, $\mathfrak{M}_{\mu=1}(U) = 1$. We now prove some fundamental results concerning nondegenerate orbits.

<u>Theorem 37.1</u>. If Γ is nondegenerate, there is a neighborhood V of Γ such that $V \setminus \Gamma$ has no ω-periodic orbits.

Proof: If $x(t) = p(t) + z(t)$ in (37.1) then

$$\dot{z}(t) = L(t,z_t) + N(t,z_t) \tag{37.4}$$

where $L(t,\varphi)$ is given in (37.3) and $N(t+\omega,\varphi) = N(t,\varphi)$ for all t,φ

$$N(t,\varphi) = f(p_t+\varphi) - f(p_t) - L(t,\varphi). \tag{37.5}$$

Consequently, $N(t,0) = 0$ and $N(t,\varphi)$ has a continuous Frechét derivative $N'_\varphi(t,\varphi)$ in φ with $N'_\varphi(t,0) = 0$. Thus, for any $\varepsilon > 0$, there is a $d > 0$ such

$$|N(t,\varphi) - N(t,\psi)| < \varepsilon|\varphi - \psi| \tag{37.6}$$

if $|\varphi|, |\psi| < d$.

The variation of constants formula applied to (37.4) asserts that $z = z(\sigma,\varphi)$ is a solution of (37.4) with $z_\sigma(\sigma,\varphi) = \varphi$ if and only if z satisfies

$$z_t = T(t,\sigma)\varphi + \int_\sigma^t T(t,s)X_0N(s,z_s)ds, \qquad t \geq \sigma, \tag{37.7}$$

where the linear operator $T(t,\sigma)\colon C \to C$ is defined by $T(t,\sigma)\psi = y_t(\sigma,\psi)$ with $y(\sigma,\psi)$ being the solution of (37.2), $y_\sigma(\sigma,\psi) = \psi$.

An element ψ of C is the initial value of an ω-periodic solution of (37.1) if and only if $\psi = p_0 + \varphi$ where φ is the initial value of an ω-periodic solution of (37.4). Therefore, from (37.6) and the periodicity of the equation, this latter statement is equivalent to saying that φ must satisfy

$$(I-U)\varphi = \int_0^\omega T(\omega,s)X_0N(s,z_s)ds \tag{37.8}$$

where z_s satisfies (37.6) and $U = T(\omega,0)$.

Suppose C is decomposed relative to the multiplier $\mu = 1$ of (37.2) as

$C = E \oplus K$ where $E = E_1(0)$, $K = K_1(0)$ are the sets defined in Section 35. Suppose $P: C \to C$ is the projection induced by this decomposition which takes C onto K and let M be a bounded right inverse of $I - U$ with $(I-P)M = 0$. If $\varphi = \varphi^E + \varphi^K$ with $\varphi^E \in E$, $\varphi^K \in K$, then the facts that $U = T(\omega,0)$, $T(\omega,0)E \subset E$, $T(\omega,0)K \subset K$, imply that

$$P(I-U)\varphi = (I-U)\varphi^K$$
$$(I-P)(I-U)\varphi = (I-U)\varphi^E = 0.$$

Therefore, equation (37.7) has a solution if and only if

$$\varphi^K = MP \int_0^\omega T(\omega,s)X_0 N(s,z_s(0,\varphi))ds$$

(37.9)

$$(I-P) \int_0^\omega T(\omega,s)X_0 N(s,z_s(0,\varphi))ds = 0$$

where $z_s(0,\varphi)$ satisfies (37.6) for $\sigma = 0$.

For any real number δ, let $B_\delta = \{\varphi \in C: |\varphi| \le \delta\}$. There is an $\alpha > 0$ such that the solutions $z(\sigma,\mu)$ of (37.4) satisfy $z(\sigma,0) = 0$ and

$$|z_t(\sigma,\varphi) - z_t(\sigma,\bar{\varphi})| \le e^{\alpha(t-\sigma)}|\varphi - \bar{\varphi}|$$

for all $t \ge \sigma$ for which $z_t(\sigma,\varphi), z_t(\sigma,\bar{\varphi}) \in B_1$. Therefore, there is a $d > 0$ such that $z_t(\sigma,\varphi) \in B_1$, $|z_t(\sigma,\varphi) - z_t(\sigma,\bar{\varphi}) \le e^{\alpha\omega}|\varphi - \psi|$ for $0 \le t - \sigma \le \omega$, $\varphi,\bar{\varphi} \in B_d$.

Now suppose that $|MP| = \beta$, $|T(\omega,s)X_0| \le \gamma$, $0 \le s \le \omega$, and ε is chosen so that $\varepsilon\beta\gamma\omega e^{\alpha\omega} < 1/2$. Let $c < 1$ be chosen so that (37.6) is satisfied for $\varphi, \psi \in C_c$. Chosen $b < d$ so that $z_t(\sigma,\varphi) \in B_c$ for $0 \le t - \sigma \le \omega$, $\varphi \in B_b$, fix $\varphi^E \in C_{b/2} \cap E$ and consider the set $\mathscr{S} = K \cap B_{b/2}$. For any $\varphi^K \in \mathscr{S}$, consider the map $\mathscr{F}: \mathscr{S} \to K$ defined by

$$\mathscr{F}\varphi^K = MP\int_0^\omega T(\omega,s)X_0N(s,z_s(0,\varphi^E+\varphi^K))ds.$$

Since $|\varphi^E+\varphi^K| \leq b$, it follows that $|z_s(0,\varphi^E+\varphi^K)| \leq c$, $0 \leq s \leq \omega$, and therefore

$$\begin{aligned}|\mathscr{F}\varphi^K| &\leq |MP| \int_0^\omega |T(\omega,s)X_0|\cdot|N(s,z_s(0,\varphi^E+\varphi^K))|ds \\ &\leq \beta\gamma\varepsilon \int_0^\omega |z_s(0,\varphi^E+\varphi^K)|ds \\ &\leq \beta\gamma\varepsilon' \omega e^{\alpha\omega}|\varphi^E+\varphi^K| \leq b/2,\end{aligned}$$

and $\mathscr{F}\colon \mathscr{S} \to \mathscr{S}$. In addition,

$$\begin{aligned}|\mathscr{F}\varphi^K-\mathscr{F}\bar{\varphi}^K| &\leq \beta\gamma\varepsilon \int_0^\omega |z_s(0,\varphi^E+\varphi^K) - z_s(0,\varphi^E+\bar{\varphi}^K)|ds \\ &\leq \beta\gamma\varepsilon\ e^{\alpha\omega}\omega|\varphi^K-\bar{\varphi}^K| < \frac{1}{2}|\varphi^K-\bar{\varphi}^K|\end{aligned}$$

and $\mathscr{F}$ is a contraction. Therefore, $\mathscr{F}$ has a unique fixed point $\varphi_*^K(\varphi^E)$ in $\mathscr{S}$ which implies that equation (37.8a) has a unique solution $\varphi_*^K(\varphi^E)$ in $K \cap B_{b/2}$. Obviously, the fixed point depends continuously upon φ^E and $\varphi_*^K(0) = 0$. Therefore, equations (37.8) have a unique solution in K given by $\varphi = 0$.

Since (37.1) is autonomous, $p(t+\alpha)$ for any real α is also an ω-periodic solution of (37.1) and the orbit of this solution is also Γ. One can, therefore, form the linear variational equation relative to $p(t+\alpha)$. This equation is

$$\dot{y}(t) = L(t+\alpha,y_t)$$

and C can be decomposed relative to the multiplier $\mu = 1$ as $C = E(\alpha) \oplus K(\alpha)$, where $E(\alpha) = E_1(\alpha)$, $K(\alpha) = K_1(\alpha)$ are the sets defined in Section 35. One can see that the projection $P(\alpha)$ and $M(\alpha)$ satisfy $|M(\alpha)P(\alpha)| \leq \beta$ for some $\beta \geq 0$. Exactly as before, one arrives at the conclusion that there is no ω-periodic solution of (37.1) with initial value $\psi = p_\alpha + \varphi$ with φ in $K(\alpha)$, except for $\varphi = 0$.

We next show there is a neighborhood U of p_0 such that for any η in U, there are unique real numbers α and φ in $K(\alpha)$ such that the function

$$G(\varphi,\alpha,\eta) = p_\alpha + \varphi - \eta = 0$$

is zero. We have $G(0,0,p_0) = 0$ and the derivative with respect to α,φ evaluated at $\alpha = 0$, $\varphi = 0$ and the pair $(\mu,\psi),\mu$ scalar, ψ in $K(0)$ is $\dot{p}_0\mu + \psi$. Since $\dot{p}_0$ is a basis for $E(0)$ and $E(0),K(0)$ are linearly independent, it is clear that this derivative has a bounded inverse. The implicit function theorem implies there is a $\delta > 0$ and unique $\alpha(\eta),\varphi(\eta)$ continuous with respect to η for $|\eta-p_0| < \delta$ so that $G(\varphi(\eta),\alpha(\eta),\eta) = 0$.

Since Γ is compact, the above argument can be applied a finite number of times to conclude that there is a neighborhood W of Γ such that any ψ in W must lie on one of the sets $[p_\alpha + K(\alpha)] \cap W$.

But the previous argument shows that one can further restrict the neighborhood W so that no ω-periodic orbits can lie on $[p_\alpha + K(\alpha)] \cap W$. This proves the theorem.

Theorem 37.2. Suppose $\omega > r$, the orbit Γ generated by the periodic solution p of (37.1) is nondegenerate and the n-vector function $G(\varphi,\varepsilon)$ is continuous in φ,ε and continuously differentiable in φ for φ in $C, 0 \leq |\varepsilon| \leq \varepsilon_0$, and $G(\varphi,0) = f(\varphi)$. Then there is an $\varepsilon_1 > 0$ and a neighborhood W of Γ such that

$$\dot{x}(t) = G(x_t,\varepsilon) \tag{37.10}$$

has a nondegenerate periodic orbit Γ_ε in W of period $\omega(\varepsilon)$, $0 \leq |\varepsilon| \leq \varepsilon_1$, $\Gamma_\varepsilon,\omega(\varepsilon)$ depend continuously upon ε, $\Gamma_0 = \Gamma$, $\omega(0) = \omega$, and Γ_ε is the only periodic orbit in W whose period approaches ω as $\varepsilon \to 0$.

Proof: For any real number $\beta > -1$, consider the transformation $t = (1+\beta)\tau$ in

(37.6). If $x(t) = y(\tau)$, then $x(t+\theta) = y(\tau+\theta/(1+\beta))$, $-r \leq \theta \leq 0$. Let us define $y_{\tau,\beta}$ as an element of the space $C([-r,0],R^n)$ given by $y_{\tau,\beta}(\theta) = y(\tau+\theta/(1+\beta))$, $-r \leq \zeta \leq 0$. Equation (37.9) becomes

$$\frac{dy(\tau)}{d\tau} = (1+\beta)G(y_{\tau,\beta},\varepsilon) \tag{37.11}$$

If (37.10) has a periodic solution of period ω, then (37.9) has a periodic solution of period $(1+\beta)\omega$ and conversely. If $y(\tau) = p(\tau) + z(\tau)$ in (37.10), then

$$\frac{dz(\tau)}{d\tau} = L(\tau,z_{\tau,0}) + H(\tau,z,\varepsilon,\beta), \tag{37.12}$$

$$H(\tau,z,\varepsilon,\beta) = N(\tau,z_{\tau,0}) + (1+\beta)G(p_{\tau,\beta}+z_{\tau,\beta},\varepsilon) - f(p_{\tau,0}+z_{\tau,0})$$

where N is defined in (37.5).

To obtain a solution of (37.11), one needs an initial function on the space $C(\beta)$ consisting of the space of initial functions mapping $[-r_0(\beta),0]$ into E^n, where $r_0(\beta) = \max\,[r,r/(1+\beta)]$. Choose β so small that $\omega > r_0(\beta)$. Let Ω_0 be the set of continuous ω-periodic functions in E^n with $\|z\|_0 = \sup_t |z(t)|$ for z in Ω_0. Lemma 36.3 implies that the nonhomogeneous linear equation

$$\frac{dz(\tau)}{d\tau} = L(\tau,z_{\tau,0}) + h(\tau), \qquad h \text{ in } \Omega_0,$$

has a solution in Ω_0 if and only if

$$\int_0^\omega q(\tau)h(\tau)d\tau = 0$$

where $q(\tau)$ is a basis for the ω-periodic solutions of the equation adjoint to (37.2). Also, since $q(\tau) \not\equiv 0$ it follows that the function q can be chosen so that $\int_0^{\omega} q(\tau)q'(\tau)d\tau = 1$ where q' is the transpose of q. For any h in Ω_0, let $\gamma(h) = \int_0^{\omega} q(\tau)h(\tau)d\tau$. Then $\gamma: \Omega_0 \to R$ is a continuous linear mapping.

For any h in Ω_0, the equation

$$\frac{dz(\tau)}{d\tau} = L(\tau, z_{\tau,0}) + h(\tau) - \gamma(h)q'(\tau)$$

has a solution in Ω_0 and a unique solution whose (I-P)-projection is zero, where P is the operator used in (37.8). If we designate this solution by $\mathscr{K}h$, then $\mathscr{K}h$ is a continuous linear operator taking Ω_0 into Ω_0.

For any positive numbers $\varepsilon_1, \beta_1, \delta_1$, let $\Omega_0(\delta_1) = \{u \text{ in } \Omega_0: \|u\| \leq \delta_1\}$ and define a map $T: \Omega_0(\delta_1) \to \Omega_0$ by the relation

$$Tu = \varepsilon\mathscr{K}[H(\cdot,u,\varepsilon,\beta) - \gamma(H(\cdot,u,\varepsilon,\beta))q'(\cdot)].$$

Using the contraction principle, one easily shows there are $\varepsilon_1, \beta_1, \delta_1$ sufficiently small so that the operator T has a unique fixed point $u^*(\varepsilon,\beta)$ in $\Omega_0(\delta_1)$, $u^*(\varepsilon,\beta)$ is a continuous function of ε, $|\varepsilon| \leq \varepsilon_1$, $|\beta| \leq \beta_1$ and $u^*(0,0) = 0$. The function $u^*(\varepsilon,\beta)$ is a solution of the equation

$$\frac{dz(\tau)}{d\tau} = L(\tau, z_{\tau,0}) + H(\tau,z,\varepsilon,\beta) - B(\varepsilon,\beta)q'(\tau) \tag{37.13}$$

where we have put

$$B(\varepsilon,\beta) = \int_0^{\omega} q(\tau)H(\tau,u^*(\varepsilon,\beta))d\tau.$$

Therefore, $u^*(\varepsilon,\beta)(t)$ is continuously differentiable in t. Using this fact and the form of H, one can reapply the contraction principle to show that $u^*(\varepsilon,\beta)$ has a continuous first derivative with respect to β. In fact, one shows that $\partial u^*(\varepsilon,\beta)/\partial\beta$ is an ω-periodic solution of the equation

$$\dot{v}(\tau) = L(\tau, v_{\tau,0}) + L_1(\tau,v,\varepsilon,\beta) + L_2(\tau,\varepsilon,\beta) - \frac{\partial B(\varepsilon,\beta)}{\partial\beta} q'(\tau)$$

where

$$L_1(\tau,v,\varepsilon,\beta) = (1+\beta)G'_\varphi(w^*_{\tau,\beta}(\varepsilon,\beta))v_{\tau,\beta} - f'_\varphi(p_{\tau,0})v_{\tau,0}$$

$$L_2(\tau,\varepsilon,\beta) = G(w^*_{\tau,\beta}(\varepsilon,\beta),\varepsilon) - \frac{1}{1+\beta} G'_\varphi(w^*_{\tau,\beta}(\varepsilon,\beta),\varepsilon)$$

$$\cdot \ [(\cdot)\dot{u}^*_{\tau,\beta}(\varepsilon,\beta) + (\cdot)\dot{p}_{\tau,\beta}]$$

and $w^*(\varepsilon,\beta) = u^*(\varepsilon,\beta) + p$. Since $\dot{u}^*_{\tau,\beta}(\varepsilon,\beta) \to 0$ as $\varepsilon \to 0$, $\beta \to 0$, it follows that $L_1(\tau,v,0,0) = 0$ and $L_2(\tau,0,0) = J(t,\dot{p})$ where $J(t,\dot{p})$ is defined in Lemma 37.2. Since we know this equation has the ω-periodic solution $\partial u^*(\varepsilon,\beta)/\partial\beta$, we must have

$$\frac{\partial B(\varepsilon,\beta)}{\partial\beta} = \int_0^\omega q(\tau)[L_1(\tau,v,\varepsilon,\beta) + L_2(\tau,\varepsilon,\beta)]d\tau.$$

From the properties of L_1, L_2 and Lemma 37.2, we have $\partial B(\varepsilon,\beta)/\partial\beta \neq 0$ for $\varepsilon = 0$, $\beta = 0$. Since $B(0,0) = 0$, the implicit function theorem implies the existence of a positive $\varepsilon_2 \leq \varepsilon_1$ and a continuous function $\beta(\varepsilon)$, $|\beta(\varepsilon)| \leq \beta_1$, $|\varepsilon| \leq \varepsilon_2$ so that $\beta(0) = 0$ and $B(\varepsilon,\beta(\varepsilon)) = 0$. Since $u^*(\varepsilon,\beta)$ is a solution of (37.12), it follows that $u^*(\varepsilon,\beta(\varepsilon))$ is an ω-periodic solution of (37.11). This proves the existence of a periodic solution $y^*(\varepsilon)$ of (37.10) of period ω and thus a solution $x^*(\varepsilon)$ of (37.10) of period $\omega(\varepsilon) = 1 + \beta(\varepsilon)$, which is continuous in ε for $0 \leq |\varepsilon| \leq \varepsilon_2$, $y^*(0) = p$. The linear variational equation associated with this periodic solution $y^*(\varepsilon)$ is a continuous function of ε and, therefore, the multiplier one will have a generalized eigenspace of dimension one for $0 \leq |\varepsilon| \leq \varepsilon_3 \leq \varepsilon_2$. The conditions of Theorem 37.1 are satisfied and there is a neighborhood W_ε of the orbit Γ_ε generated by $y^*(\varepsilon)$, $0 \leq |\varepsilon| \leq \varepsilon_4 \leq \varepsilon_3$ such that the equation (37.10) has no ω-periodic orbit in $W_\varepsilon \backslash \Gamma_\varepsilon$. The proof of Theorem 37.1 also shows that one can choose W independent of ε for $0 \leq |\varepsilon| \leq \varepsilon_4$. This proves the theorem.

38. NOTES AND REMARKS

§2-5. For a treatment of related questions for neutral equations and more general retarded equations, see Driver [3], Cruz and Hale [1], Hale [8], Melvin [1], Jones [2].

§6. One can extend the definition of atomic at $-r$ to the case in which the function $f(t,\varphi)$ is only Lipschitzian in φ. For a suggestion on how this may be done, see Hale and Cruz [1]. The results on backward continuation then overlap with Hastings [1]. Hastings has other interesting results on the density of the initial values of those solutions of a linear equation which are defined on $(-\infty,0]$. Backward continuation for neutral equations is discussed by Hale [8].

§8. Krasovskii was the first to emphasize the importance of considering the state of a system defined by a functional differential equation as the element $x_t(\sigma,\varphi)$ of C. He made the observation that the converse theorems of Lyapunov on stability (see Section 11) could not be proved by using a scalar function $V(t,x)$ which depends only upon (t,x) in $R \times R^n$. In fact, if uniform asymptotic stability of the solution $x = 0$ of

$$\dot{x}(t) = f(x_t)$$

implies the existence of a positive definite function $V(x)$ such that $(\partial V/\partial x)f < 0$, then the solution $x = 0$ of

$$\dot{x}(t) = kf(x_t)$$

would be uniformly asymptotically stable for any positive k since $(\partial V/\partial x)(kf) < 0$. On the other hand, the linear equation

$$\dot{x}(t) = -kx(t-1)$$

has all roots of the characteristic equation $\lambda = -ke^{-\lambda}$ with negative real parts if $k < \pi/2$ and some with positive real parts if $k > \pi/2$.

The example of Remark 8.12 is due to Zverkin [1] and the one of Remark 8.13 to Yorke and Winston [1].

§9. Invariant sets for functional differential equations were first mentioned explicitly by Hale [1].

§10. Lemma 10.2 is due to Zverkin [1].

§11. Krasovskii [2, pp. 151ff] proved asymptotic stability under the hypotheses of Theorem 11.1. The proof in the text was communicated to the author by Yoshizawa (see also the book of Yoshizawa [1]). The conclusion of asymptotic stability of Theorem 11.2 is contained in Krasovskii [2, p. 157ff], Razumikhin [1]. The uniform asymptotic stability was proved by Driver [1]. Example 11.3 is due to Krasovskii [1, p. 174]. For Liapunov functions and neutral equations, see Cruz and Hale [1], Infante and Slemrod [1], Slemrod [1].

§12. A special case of Theorem 12.1 was given by Shimanov [1].

§13. This material is based on Hale [2] taking into account the improvements by LaSalle [1].

§14-15. These examples are based on Hale [2]. Many more examples are contained in that paper.

§16. The derivativations in this section follow the ones of Hale and Meyer [1] for neutral equations.

§17. The adjoint equation has been used in functional differential equations since 1920. For a complete list of references on its evolution, see Zverkin [4].

§19-24. These sections on autonomous linear systems are based on a paper by Hale [3] and follows the presentation in Hale and Meyer [1]. For the retarded functional differential equations, the presentation could be shortened some, but the approach in the text is taken because very little change (except for §22) is needed to discuss neutral equations. The estimates on the complementary subspaces are very difficult and have only recently been obtained in a precise manner by D. Henry [2].

§25. The results of this section are immediately applicable to systems of the form

$$\dot{x}(t) = L(t,x_t) + \varepsilon g(t,x_t,\varepsilon)$$

where ε is a small parameter. For example, if the equation (25.2) has a unique solution in $\mathscr{B}$, for every $f \in \mathscr{B}$, then Theorem 25.1 implies the existence of a bounded linear operator $\mathscr{K}: \mathscr{B} \to \mathscr{B}$ such that $\mathscr{K}f$ is the unique solution of (25.2). Therefore, the above equation has a solution in $\mathscr{B}$ if and only if x satisfies

$$x_t = \varepsilon(\mathscr{K}g(\cdot,x,\varepsilon))_t.$$

One can now apply the usual fixed point principles to obtain the existence of solutions in $\mathscr{B}$.

If we consider perturbations of autonomous linear systems

$$\dot{x}(t) = L(x_t) + \varepsilon g(t,x_t,\varepsilon),$$

then one can do even more. If the linear equation

$$\dot{x}(t) = L(x_t)$$

has some eigenvalues on the imaginary axis and the set of these eigenvalues are denoted by Λ_0, then C can be decomposed as in Section 24 as $P \oplus Q$, $P = P_{\Lambda_0}$ to obtain an equivalent set of equations

$$x_t = \Phi y(t) + x_t^Q$$

$$\dot{y}(t) = By(t) + \varepsilon g(t,x_t,\varepsilon)$$

$$x_t^Q = T(t-\sigma)x_\sigma^Q + \varepsilon \int_\sigma^t T(t-s)X_0^Q g(s,x_s,\varepsilon)ds.$$

One can now generalize the usual method of obtaining periodic and almost periodic solutions of such equations. We do not dwell on this point, but simply give the references for the results. For periodic solutions, see Perello [1], for the methods of averaging, see Hale [4], Tolosa [1], and for an interesting bifurcation problem see Chafee [1]. Using the above decomposition, Hale [5], Cooke [2], Kato [1] have discussed the asymptotic behavior of solutions of linear systems which are close to autonomous ones. The stability in critical cases for retarded and neutral equations is discussed in Hale [9].

§26. Much more detailed information on the behavior of perturbed linear systems (even nonautonomous ones) may be found in Hale and Perello [1]. For the saddle point property for neutral equations, see Cruz and Hale [2].

§27. Theorem 27.1 is due to Krasnoselskii [1]. Theorems 27.2 and 27.3 are due to Grafton [1].

§28. Theorem 28.1 is due to Grafton [1].

§29. Lemma 29.1 is due to Wright [1]. Theorem 29.1 was first proved by G. Stephen Jones [1]. The proof in the text follows very closely the one of Grafton [1]. The proof of Jones used very detailed information about the expansions of the solutions of (29.1) in terms of the characteristic functions of the linear equation

$$(29.3) \qquad \dot{x}(t) = -\alpha x(t-1)$$

and an asymptotic fixed point theorem of F. E. Browder. The general theorem of Section 28 was not too easy to prove, but the application of this result to the above example requires very little information about the characteristic equation of (29.3) and no information about expansions of solutions.

Jones [3,4] has obtained other fixed point theorems which are applicable for asserting the existence of periodic solutions of functional differential equations and do not use expansion theorems for the solutions.

§30. Theorem 30.1 was first proved by Jones [1] and the proof of the text follows Grafton [1].

§31. The proof of Theorem 31.1 is based on the one given by Grafton [1] for the equation $\ddot{x}(t) - k[1-x^2(t)]x(t) + x(t-r) = 0$. The proof of the lemma on the zeros of the characteristic polynomial is due to D. Henry.

§32. The general representation of solutions (Theorem 32.2) was first proved by Banks [1]. For the representation theorem for neutral equations, see Henry [4].

§33. Henry [3] was the first to study the function space adjoint of retarded functional differential equations. For the more general treatment for neutral equations, see Henry [4].

§34. This general treatment of boundary value problems is based on Henry [3]. Corollary 34.1 was first proved by Halanay [1] and the special boundary value problem (34.12), (34.13) under the assumption that M, N commute with $A(t)$ is due to Halanay [3]. For a different treatment of Corollary 34.1, see Wexler [1]. For a different treatment of boundary value problems, see Bancroft [1]. For the neutral case, see Henry [4].

§35. The paper of Stokes [1] was the first general discussion of the Floquet theory for periodic functional differential equations. Theorem 32.1 is due to Stokes [1]. The case of differential difference equations with integer lags has received much attention. In particular, see Hahn [1], Zverkin [2,3], Lillo [1,2] for the difficulties involved in trying to obtain expansion theorems in terms of the Floquet solutions. It is tempting to conjecture that there may be a periodic transformation of variables in C which would reduce the periodic functional differential to an autonomous one. Henry [1] has shown that any solution of an autonomous equation which approaches zero faster than an exponential must be identically zero after some time. The example (35.6) therefore, shows that such a transformation of variables is impossible for the general periodic system.

§36. Shimanov [2] was the first to state the decomposition theorem for periodic systems for the special case when the function $\eta(t,\theta)$ as a function of bounded variation in θ has no singular part. The presentation in the text follows Henry [3].

§37. The material of this section is based on Hale [6]. If Γ is a non-degenerate periodic orbit and all characteristic multipliers $\mu \neq 1$ satisfy $|\mu| \neq 1$, then some results have been obtained concerning the saddle point structure near Γ (see Hale [6]). The analogue of the orbital stability theorem of Poincaré has been given by Stokes [2].

BIBLIOGRAPHY

Trudy Seminara Po Teorii Differentialnix Yravneniya C Otklonyayoshchimsya Argumentom, Vol. 1(1962), Vol. 2(1963), Vol. 3(1965), Vol. 4(1967), Vol. 5(1967), Vol. 6(1968), Vol. 7(1969), Univ. Dryzbi Narodov im. Patrisa Lymumbi, Moscow. (Devoted entirely to hereditary equations.)

Banks, H.T., [1] The representation of solutions of linear functional differential equations, J. Differential Equations, 5(1969), 399-410.

Bancroft, S., [1] Boundary value problems for functional differential equations, Ph.D. Thesis, Brown University, Providence, R.I. (1968).

Bellman, R. and K. Cooke, [1] Differential-Difference Equations, Academic Press, 1963.

Bellman, R., [2] Asymptotic behavior of solutions of differential-difference equations, Memoirs American Mathematical Society, 35(1959).

Brayton, R., [1] Bifurcation of periodic solutions in a nonlinear difference-differential equation, IBM Research paper, RC-1427, June 1965.

Brownell, F. H., [1] Nonlinear differential-difference equations, Contr. to Theory of Nonlinear Oscillations, Vol. 1, Annals of Mathematical Studies, 20(1950), Princeton.

Chafee, N., [1] Bifurcation problems, Journal of Mathematical Analysis and Applications. To appear in 1971.

Choksy, N. H., [1] Time lag systems - A bibliography, IRE Transactions on Automatic Control, Vol. AC-5, January 1960. Supplement 1, published somewhere (probably same journal). (A good bibliography.)

Cooke, K., [1] The condition of regular degeneration for singularly perturbed linear differential-difference equations, Journal of Differential Equations, 1(1965).

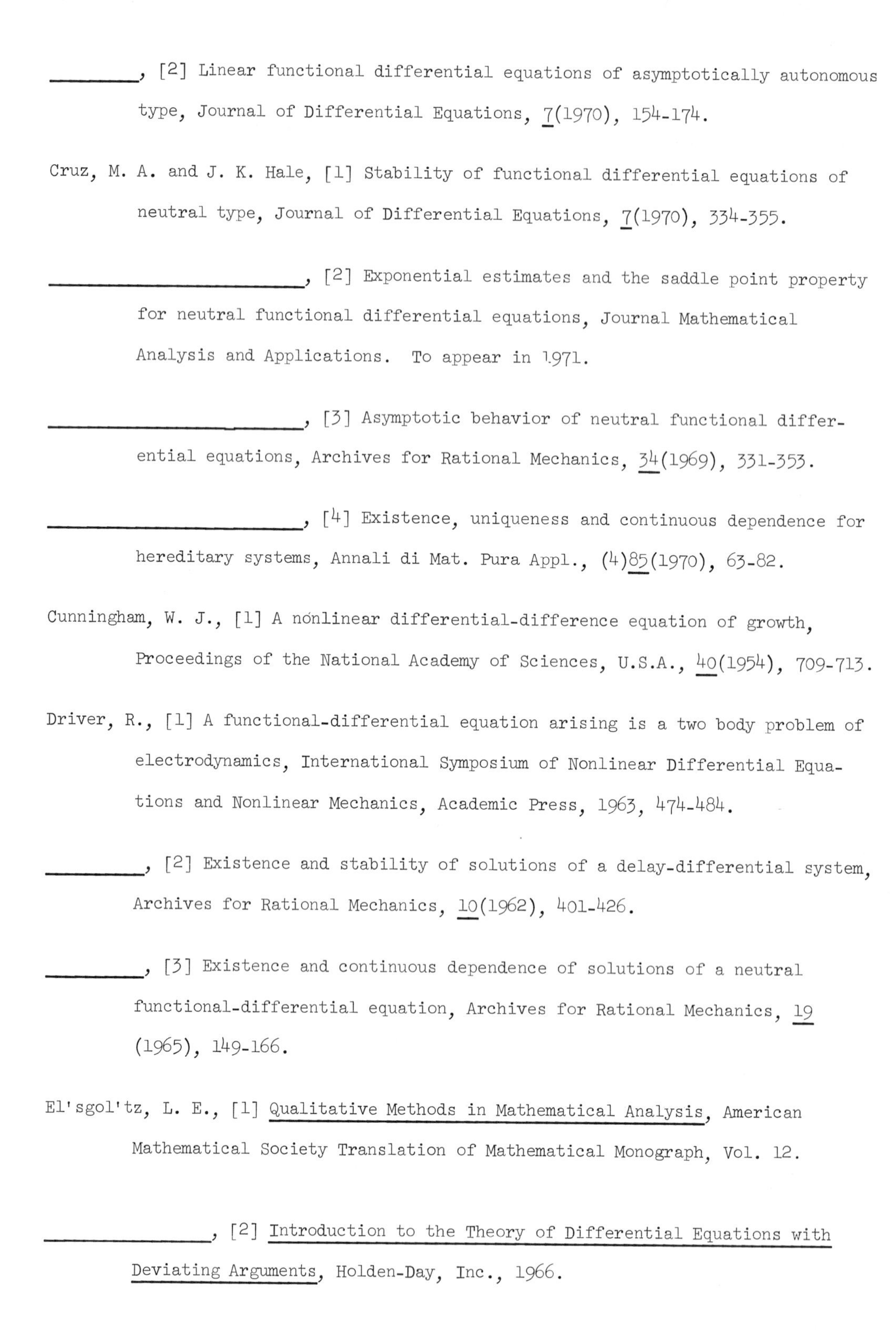

________, [2] Linear functional differential equations of asymptotically autonomous type, Journal of Differential Equations, 7(1970), 154-174.

Cruz, M. A. and J. K. Hale, [1] Stability of functional differential equations of neutral type, Journal of Differential Equations, 7(1970), 334-355.

________________________, [2] Exponential estimates and the saddle point property for neutral functional differential equations, Journal Mathematical Analysis and Applications. To appear in 1971.

________________________, [3] Asymptotic behavior of neutral functional differential equations, Archives for Rational Mechanics, 34(1969), 331-353.

________________________, [4] Existence, uniqueness and continuous dependence for hereditary systems, Annali di Mat. Pura Appl., (4)85(1970), 63-82.

Cunningham, W. J., [1] A nonlinear differential-difference equation of growth, Proceedings of the National Academy of Sciences, U.S.A., 40(1954), 709-713.

Driver, R., [1] A functional-differential equation arising is a two body problem of electrodynamics, International Symposium of Nonlinear Differential Equations and Nonlinear Mechanics, Academic Press, 1963, 474-484.

________, [2] Existence and stability of solutions of a delay-differential system, Archives for Rational Mechanics, 10(1962), 401-426.

________, [3] Existence and continuous dependence of solutions of a neutral functional-differential equation, Archives for Rational Mechanics, 19 (1965), 149-166.

El'sgol'tz, L. E., [1] Qualitative Methods in Mathematical Analysis, American Mathematical Society Translation of Mathematical Monograph, Vol. 12.

______________, [2] Introduction to the Theory of Differential Equations with Deviating Arguments, Holden-Day, Inc., 1966.

Ergen, W. K., [1] Kinetics of the circulating fuel nuclear reactor, Journal of Applied Physics, 25(1954), 702-711.

Grafton, R., [1] A periodicity theorem for autonomous functional differential equations, Journal of Differential Equations, 6(1969), 87-109.

Hahn, W., [1] On difference differential equations with periodic coefficients, J. Mathematical Analysis and Applications, 3(1961), 70-101.

________, [2] Zur stabilitat der losungen von linearen differential-diffenzengeichungen mit knostanten koeffizienten, Math. Annalen, 131(1965), 151-166.

________, [3] Theory and Application of Lyapunov's Direct Method, [1] Prentice-Hall, 1963.

Halanay, A., [1] Teoria Calitativa A Ecuatiilor Diferentiale, Editura Acad. Rep. Populare Romine, 1963. Translated by Academic Press, 1966 as Differential Equations.

________, [2] Systems a retardement. Resultats et problemes, Third Conference on Nonlinear Vibrations, E. Berlin. Akademie-Verlag, Berlin, 1965. (Survey Article)

________, [3] A boundary value problem for linear systems with time lag, Journal of Differential Equations, 2(1966), 47-56.

________, [4] The method of averaging in equations with retardation, Revue de Mathematiques Pures et Appliquees, 4(1959), 467-483.

Hale, J. K., [1] A stability theorem for functional-differential equations, PNAS 50(1963), 942-946.

________, [2] Sufficient conditions for stability and instability of autonomous functional-differential equations, Journal of Differential Equations, 1(1965), 452-482.

__________, [3] Linear functional-differential equations with constant coefficients, Cont. Differential Equations, 2(1963), 291-319.

__________, [4] Averaging methods for differential equations with retarded arguments and a small parameter, Journal of Differential Equations, 2 (1966), 57-73.

__________, [5] Linear asymptotically autonomous functional differential equations, Rendiconti del Circolo Matematico di Palermo, (2)15(1966), 331-351.

__________, [6] Solutions near simple periodic orbits of functional differential equations, Journal of Differential Equations, 9(1970), 126-138.

__________, [7] A class of functional-differential equations, Cont. Differential Equations, 1(1963), 411-423.

__________, [8] Forward and backward continuation for neutral functional differential equations, Journal of Differential Equations, 9(1971).

__________, [9] Critical cases for neutral functional differential equations, Journal of Differential Equations, 9(1971).

Hale, J. K. and K. Meyer, [1] A class of functional differential equations of neutral type, Memoirs of the American Mathematical Society, No. 76, 1967.

Hale, J. K. and C. Perello, [1] The neighborhood of a singular point of functional-differential equations, Cont. Differential Equations, 3(1964), 351-375.

Henry, D., [1] Small solutions of linear autonomous functional differential equations, Journal of Differential Equations, 8(1970), 494-501.

__________, [2] Estimates on the complementary subspaces of neutral equations. To appear.

__________, [3] The adjoint of a linear functional differential equation and boundary value problems, Journal of Differential Equations, 9(1971).

________, [4] Adjoint theory and boundary value problems for neutral functional differential equations. To appear.

Hastings, S. P., [1] Backward existence and uniqueness for retarded functional differential equations, Journal of Differential Equations, 5(1969), 441-451.

Hughes, D. K., [1] Variational and optimal control problems with delayed argument, Journal of Optimization Theory and Applications, 2(1968), 1-14.

Infante, E. F. and M. Slemrod, [1] Asymptotic stability for linear systems of differential-difference equations of neutral type and their discrete analogues. To appear in Journal of Mathematical Analysis and Applications, 1971.

Jones, G. S., [1] The existence of periodic solutions of $f'(x) = -\alpha f(x-1)\{1+f(x)\}$, Journal of Mathematical Analysis and Applications, 5(1962), 435-450.

________, [2] Hereditary dependence in the theory of differential equations II, University of Maryland Technical Report, 1965.

________, [3] Periodic functions generalized as solutions of nonlinear differential-difference equations, International Symposium of Differential Equations and Nonlinear Mechanics, Academic Press, 1963, 105-112.

________, [4] Asymptotic fixed point theorems and periodic solutions of functional differential equations, Cont. Differential Equations, 2(1963), 385-405.

Kato, J., [1] On the existence of a solution approaching zero for functional differential equations, Proceedings U.S.-Japan Seminar Differential and Functional Equations, 153-169, W. A. Benjamin, 1967.

Krasnoselskii, M. A., [1] Postive Solutions of Operator Equations, P. Noordhoff, Ltd., 1964, p. 561.

Krasovskii, N., [1] On the stabilization of unstable motions by additional forces when the feedback loop is incomplete, Prikl. Mat. Mek., 27(1963), 641-663; TPMM, 971-1004.

____________, [2] Stability of Motion, Moscow, 1959, Translation, Stanford University Press, 1963.

LaSalle, J. P., [1] An invariance principle in the theory of stability, International Symposium on Differential Equations and Dynamical Systems, Academic Press, 1967, p. 277.

Levin, J. J. and J. Nohel, [1] On a nonlinear delay equation, Journal Mathematical Analysis and Applications, 8(1964), 31-44.

Lillo, J. C., [1] Periodic differential difference equations, Journal Mathematical Analysis and Applications, 15(1966), 434-441.

____________, [2] The Greens' function for periodic differential difference equations, Journal of Differential Equations, 4(1968), 373-385.

Marchenko, Y. I. and V. P. Rubanik, [1] On mutual synchronization of molecular generators, Izv. Vis. Uch. Zav. Radiophsica, 8(1965), 679-687.

Melvin, W., [1] A class of neutral functional differential equations, Ph.D. Thesis, Brown University, Providence, Rhode Island, 1971.

Miller, R. K., [1] Asymptotic behavior of nonlinear delay-differential equations, Journal of Differential Equations, 1(1965), 293-305.

Minorsky, N., [1] Nonlinear Oscillations, D. Van Nostrand Company, Inc., Princeton, 1962.

Miranker, W. L., [1] Existence, uniqueness and stability of solutions of systems on nonlinear difference-differential equations, IBM Research Report, RC-322, 1960.

______________, [2] The wave equation with a nonlinear interface condition, IBM Journal Research and Development, 5(1961), 2-24.

Mishkis, A. D., [1] General theory of differential equations with a retarded argument, American Mathematical Society, Translation, 55(1951). (Survey article).

______________, [2] Lineare Differentialgleichungen Mit Nacheilenden Argument, Deutscher Verlagder Wissenschaften, Berlin, 1955.

Norkin, S. B., [1] Differential Equations of the Second Order With Retarded Arguments, (Russian) Moscow, 1965.

______________, [2] Oscillation theorems of the type of Sturm for differential equations of the second order with retarded arguments, Nauch. Dokl. Vish. Skoli, Fiz. Mat. Nauk, 2(1958), 76-80.

Perello, C., [1] Periodic solutions of differential equations with time lag containing a small parameter, Journal of Differential Equations, 4(1968), 160-175.

Pinney, E., [1] Differential-Difference Equations, University of California Press, 1958.

Pitt, H. R., [1] On a class of integro-differential equations, Proceedings of the Cambridge Philosphical Society, 40(1944), 119-211; 43(1947), 153-163.

Razumikhin, B. S., [1] On the stability of systems with a delay, Prikl. Mat. Mek., 20(1956), 500-512.

______________, [2] Application of Liapunov's method to problems in the stability of systems with a delay, Avtomat. i Telemeh., 21(1960), 740-748.

Shimanov, S. N., [1] On the instability of the motion of systems with retardations, PMM 24(1960), 55-63; TPMM 70-81.

__________, [2] On the theory of linear differential equations with periodic coefficients and time lag, Prikl. Mat. Mek., 27(1963), 450-458; TPMM, 674-687.

__________, [3] On the theory of linear differential equations with retardations, Differentzialnie Uravneniya, 1(1965), 102-116.

__________, [4] On stability in the critical case of a zero root with a time lag, Prikl. Mat. Mek., 24(1960), 447-457.

__________, [5] On the vibration theory of quasi-linear systems with time lag, Prikl. Mat. Mek., 23(1959), 836-844.

Slater, M. and H. S. Wilf, [1] A class of linear differential difference equations, Pacific Journal of Mathematics, 10(1960), 1419-1427.

Slemrod, M., [1] Nonexistence of oscillations in a nonlinear distributed network, to appear in Journal Mathematical Analysis and Applications, 1971.

Snow, W., [1] Existence, uniqueness and stability for nonlinear differential-difference equations in the neutral case, Ph.D. Thesis, New York University, 1964.

Stokes, A., [1] A floquet theory for functional-differential equations, Proceedings of the National Academy of Sciences, 48(1962), 1330-1334.

__________, [2] On the stability of a limit cycle of an autonomous functional-differential equation, Cont. Differential Equations, 3(1964), 121-140.

Tolosa, K. K., [1] The method of averaging for quasi-linear differential equations with retarded arguments and periodic coefficients, Trudy Sem. po teorii ..., 6(1968), 147-159.

Volterra, V., [1] Theorie Mathematique De La Lutte Pour La Vie, Gauthier-Villars, 1931.

__________, [2] Sulle equazioni integro-differenziali della teoria dell' elasticita, Atti Reale Accad. Lincei, 18(1909), 295.

Wangersky, P. J. and W. J. Cunningham, [1] Cold spring harbor symposium on quantitative biology, Vol. 22, Population Studies: Animal Ecology and Demography (1957); The Biological Laboratory, Cold Spring Harbor, Long Island, New York.

Wexler, D., [1] Solutions periodiques des systems lineaires a argument retarde, Journal of Differential Equations, 3(1967), 336-347.

Wright, E. M., [1] A functional equation in the Heuristic theory of primes, The Mathematical Gazette, 45(1961), 15-16.

__________, [2] A nonlinear differential difference equation, Journal Reine Augew. Math., 194(1955), 66-87.

Yoshizawa, T., [1] Stability Theory by Lyapunov's Second Method, Mathematical Society, Japan, 1966.

__________, [2] Extreme stability and almost periodic solutions of functional-differential equations, Archives for Rational Mechanics, 1965.

Zverkin, A. M., [1] Dependence of the stability of the solutions of differential equations with a delay on the choice of the initial instant, Vestnik Moskov. Univ. Ser. Mat., 5(1959), 15-20.

__________, [2] Appendix to Russian translation of Bellman and Cooke, Differential-Difference Equations, MIR, Moscow, 1967.

__________, [3] Completeness of the eigensolutions of an equation with lags and periodic coefficients, Trudy Sem. po teorii ..., 2(1963), 94-111.

__________, [4] Expansion of solutions of differential difference equations in series, Trudy Sem. po teorii ..., 4(1967), 3-50.

Zverkin, A. M., Kamenskii, G. A., Norkin, S. B. and L. E. El'sgol'tz, [1] Differential equations with retarded arguments, Uspehi Mat. Nauk, 17(1962), 77-164. (Survey article).

Rubanik, V. P., [1] Oscillations of Quasilinear Systems with Retardations, Nauka, Moscow, 1969 (Russian).

INDEX